Blindsight

A
Quantum Life
in a
Human World

Blindsight

A
Quantum Life
in a
Human World

Mikal Masters

Just because you don't see it
doesn't mean it isn't there~
— MWUSO

LEON SMITH
PUBLISHING
www.leonsmithpublishing.com

Back cover photo by: Alice Rabbit
www.AliceRabbit.com

ISBN: 978-1-945446-37-5

www.blindsight.mobi

Always for G~

Acknowledgements

Acknowledgements are all about gratitude, being thankful in a deep and heartfelt way for those who have impacted and continue to impact our life, our journey, and this project~ Here in our seventh decade of our manifestation, the list of those to whom we are indebted and grateful is simply too long~ Do we thank our Uncle Lee for teaching us at age nine how to throw a curveball without hurting our arm? What about the beautiful girl who gave us our first kiss? There was our high school English teacher who told us we should think about becoming a writer~ Another amazing teacher encouraged us to try public speaking and debate, and we went on to minor in the subject in college~ We are so grateful to the local radio station owner who mistook our application to answer the phones for him on weekends~ He gave us a job as an on-air announcer instead when we were only sixteen years old~

Undying thanks goes to the late Rusty Weir who drew us out of a local Austin music scene and brought us into the national arena of recording, songwriting, and performing~ Rusty never gave less than 110 percent on stage and showed us every night—through hundreds of shows—that no matter how you are feeling, the audience deserves simply everything~

Here's to Ernie Gammage, our old road mate, who managed to demonstrate on stage every night how to surf streams of consciousness just outside the normal with humor and energy~ Always brilliant, never doing harm, Ernie is a true musical force and a lifelong friend~

We can't leave out Lynn Robinson who found us working in a bowling alley lounge for $100 a night and tips~ She lured us to her magical home and bar in the Altamont Hills and filled our world with gypsies, Harley-Davidsons and rockin' blues~ Lynn, you forever changed our life~

The Random Acknowledgement Award goes uncontested to Don Shepherd, author of the upcoming book, Soft Landing~ A chance encounter in a mattress store led to several hours of discussion about quantum mechanics and metaphysics until the store closed and the manager politely asked us to leave~ More talks followed, along with an enduring friendship and mutual mentorship~ Through these parleys, we found the dream, and ultimately the confidence, to write this book. We feel undying gratitude beyond words~

All these people and events, along with many others, created our path and influenced us, leading us to this point~ With some regret, we will attempt to narrow the list~

The single most important person for whom we are thankful beyond meager words is G, our love mate, life partner, mother of our amazing daughters and enduring voice of reason grounding us in our ever-wandering imaginations~ She is simply the most centered and gifted human we have ever encountered~ Her gifts are never forced on anything or anyone; they simply appear when they are needed~ Sharing our lives is our single greatest treasure~

Our oldest daughter is our Millennial Light Bearer with extraordinary sensitivities that bring love and honor to disadvantaged children, many of whom had never experienced these qualities~ Simply, no one had previously seen and heard these children, and the children had never learned how to return the same~ Erin Nicole is changing the world with her remarkable gifts, and we are so grateful for her in our life~ She is also our regular sounding board for all of our Multiversal musings~ She is also the extraordinary artist featured in most of the art in oils and watercolor herein~ Thanks so much for listening and responding~

Kelci Auliya is our youngest and the most intelligent human we have ever known~ There are so many crazy-smart people out there, but true intelligence that spans across critical thinking, intuition, creativity, art, music, animals, medicine, mathematics, mechanicals, business, and metaphysics is extremely rare~ This intelligence comes in a package that works hard and plays just as hard~ She is heavily involved in the rescue of endangered and the most handicapped of dogs, working tirelessly as a foster to place them in loving homes~ With her degree in sports medicine and rehabilitation, she dreams of a rehab ranch to treat and rescue dogs~ In early renderings of our manuscript, she worked hard on our grammar and syntax while understanding better than anyone what we were attempting to communicate~ She is the artist behind our Sunbird, the Blindsight logo, the Infinite Yin Yang, the Quantum Ouroboros, and the Multiversal Guitar among others~ We are so grateful for you~

These three make up our amazing family, the best of friends and a love-brain trust for whom the infinite possibilities continue to unfold~

We are especially grateful for Keith and Maura Leon, whom we met serendipitously through our daughter, Erin Nicole~ After only a few short telephone conversations and an hour at a coffee shop outside of San Antonio, they have become fast friends, mentors and publishers~ We look forward to a long and prosperous relationship~ Here, we must include the amazing Autumn Carlton, or as Keith refers to her, my other me, who has done much to guide our process through to this finished product~

Through Keith and Maura, we were teamed with the brilliant, funny, and insightful Karen Burton. She took on the daunting task of editing the tangled manuscript of a first-time author who insisted on breaking far too many traditional rules~

With her team, she reigned in our language while preserving the heart of its meaning~ It was Karen who gently placed the idea of these Acknowledgements in our path~ The more we have gotten to know this woman, the more we are humbled by her depth of intelligence and intuition~ We are so thankful for your patience and your friendship~

Our Above-and-Beyond-the-Call Award goes to the mysterious Dot, our proofreader, who boldly accepted the challenge of learning our new language and creative phrasing~ By embracing the uniqueness of our approach, she managed to ground our manuscript and keep it reasonably accessible~ Thanks so much for your insights~

Finally, lasting thanks to Nida and her Nida's Touch for beautifully coordinating the images with Rudy, who then took our outrageous approach to layout design and brought it into palatable reality~

This page has expanded our capacity for gratitude, love, and Intellergy~ It is the work of one very fortunate son~

Contents

Buddha Mathan by Erin Nicole, age 8

I

A Little Story

~~~*~~~

Horace so loved his home~ As he swam through the warm, murky shallows, he waited for smaller, unsuspecting fish to wander by~ He felt powerful as he effortlessly sprang and snapped up his prey~ His friends, Lolly and Bazo, were never far away, as they spent their days together roving their little swamp playing, exploring and hunting~

Of course, they had no idea they would one day be called Tetrapods, living almost 370 million years ago in the Devonian Period, long before any dinosaur roamed the earth~ Horace and his little school also had no knowledge that an animal had never, ever walked on the land~

What they did have was a feeling somewhere deep inside them, like an uncomfortable pressure or a wanting that never seemed to go away~ They seemed to know there was something out there larger than themselves, just out of reach, just out of sight~ Sometimes after a storm, they would see a brightening on the surface and it was times such as these that the wanting became more intense~

They had always believed in the teachings of the Old Ones – that to go into the brightening was almost certain death or insanity — so they remained in the shadowy depths, finding safety and contentment in an existence that had gone unchanged since before any could remember~

One morning after a huge tempest, Horace felt the marsh brighten and something stirred deep inside of him~ He felt himself drawn ever upward toward the light, knowing in his heart it was dangerous~ If his parents saw him, they would call out for him to stop and return now~

As he drifted upward toward the ever-lightening glow, he heard Lolly and Bazo cry, "Horace, no! Don't!" "What do you think you're doing? You are almost to the light!"

By now, Horace could hear nothing but the calling of his heart and soul being drawn ever higher from the depths~ All of a sudden, he burst from the water and found himself out on a small sandy shore~ Sunlight blazed into his eyes and he was temporarily blinded~ Unable to breathe the life-affording waters through his ancient gills, and in mortal fear for his life, he began to struggle and fight for all his worth to return home, when the most remarkable thing happened: he took a breath of fresh air~

Lungs that had lain dormant for so long suddenly found purpose and, as his sight slowly adjusted, he saw he was now in a world of endless blue sky~ Everywhere on the sandy, little bank grew ferns and horsetails, and he knew at his core that he was in a place of wonder, not the realm of death and insanity he had always believed awaited those who sought the light~ He looked down as some sort of insect wiggled in front of him~ He instinctively lunged at this newfound quarry, finding the fresh burst of juicy flavor surprisingly delightful~

His bony fins began to move instinctively in unusual and new ways, pulling him further onto the land, and he soon found he was just as comfortable moving, breathing, and hunting here as he was at home~ His murky shallows would never be the same again~

Behind him lay the swamp, and all he could think about was to return home to tell Bazo and Lolly of his amazing adventures~ He turned and made his way into the familiar gloom, but something was different~ Horace knew that his true home was much bigger now, much brighter~ There was so much more for him to explore and learn~ He knew to the core of his being that he could no longer be just a happy lungfish in a muddy little pond~

He swam anxiously with his excitement until he finally found his friends, enthusiastically telling them of all he had seen, felt, and found~ When he had finished, he waited for their passion to join with his, but instead what he saw looked a lot more like fear~

"Come on, Horace, what really happened up there?" Lolly asked, backing away slightly~ "This just doesn't sound anything like you."

"You're talking a little crazy here, my friend," Bazo added~ "All this about a big blue, breathing air and wiggly food? I don't know what to think here."

"I'm telling you it's all true~ You have to be there yourself to know~ Please trust me~ This is where we are supposed to be~ This is what we truly are," Horace pleaded~ "I'm going back~ You can stay here if you like, but I hope to see you on the other side~ For now, goodbye."

Horace swam slowly to the surface, dazed by the responses of his dearest friends but committed to his new life~ Now that he had seen the new world, there was no way he could remain where he had been~

He crawled out of the murk, a little farther this time, sampling from bites of fern and bugs~ He rolled over onto his back and felt the warmth of the sun~ Never before had he felt warm, and the sensation was so enlightening~ He felt his mind expanding with the new possibilities and potential that awaited~ He really didn't know where to begin~

"I think I will just lie back and enjoy the beach for a while," he smiled to himself~

After a time, Horace looked back over to the shoreline~ He saw what looked like two noses peeking through the surface~

~~~*~~~

NOTES

Machdnach Photo by G~

II

Auto Bio Intro

Somewhere around sixty years ago~

The questions for us began around the age of six years when, in the span of about a year, we managed to contract virtually every childhood disease available to us at the time~ As a result, over one third of our first-grade school year was spent in bed at home, either recuperating from the likes of measles, mumps, chicken pox, scarlet fever, bronchitis, strep throat, flu, and a couple colds mixed in, or shaping a lifelong, intimate relationship with the piano~ The physical body took a crash course in immune system construction~ There has been very little illness since~

The reason this is relevant is that during the course of these events, we experienced some very high spikes in body temperature, in the 104°F to 105°F degree range~ Associated with these fevers, we had two extremely vivid out-of-body experiences~ We *floated* above our bodies, clearly watching the doctor and our parents, who stood over us, visibly concerned, while we were administered the typical penicillin dose through a hypodermic, in our minds the size of a four-inch nail, followed by a bonus in the form of a vitamin B12 injection~ The two shots per office visit were something every child dreaded~ Luckily, we were not *present* for these two~

Later, upon our *return* to the body, we told of our experiences to our parents and the doctor~ They easily dismissed them as simple *fever dreams*, but many was the time in subsequent years in the quiet just before sleep, when we would relive those reverie voyages, becoming more and more convinced they were anything but dreams~

It was only many years later when, encountering stories of the lives of certain shamans, mystics, and healers, that we understood that many of them had had similar experiences resulting from high fevers as children~ Fortunately for them, they had manifested in a sympathetic environment and had been met with nurturing, understanding, and later instruction from those who were familiar with similar experiences themselves~ In our small Texas town, we were born into a culture with no frame of reference for this type of event~ This was a path we would have to forge for ourselves~

Along with this newfound enjoyment we had found at play inside our minds, we began to notice that in the dark and silence of the time before sleep, we repeatedly saw extremely complex and vividly colorful *wheels* made up of the most fantastic geometrical figures when we looked up into our heads~ At our early age, with absolutely no fear or preconception, we began to toy with them~ As a child, we had no reason to doubt that everyone was doing this~ Soon we learned how to make them spin in rotation, and later to reverse the spin, slow them down, speed them up, and make them recede or advance toward us~ We then found we could change the colors into any we liked, along with generating amazing new patterns~ Many was the time the colors got so bright we had to open our eyes in the darkness of our room to gain relief from the intensity~

As with the discovery years later of the connection between the fevers and the shamans, so it was remarkable in later years to discover the Hindu mandala~ We can still remember the amazing familiarity, the *déjà vu*, as we looked at drawings of the very same *wheels* with which we had been playing in our heads for so long~ We have to say that our wheels have always been infinitely more complex and interactive~

At some point years later, there was a night when, in the middle of our wheel play, the figure took on the image of a full-screen human iris and pupil~ Our immediate impression was that this *eye* was self-aware and observing us~ As of this writing, we know now that the presence we felt was merely a projection of our own egos, but nevertheless it frightened us to the core, and we became afraid to close our eyes for a short time~ Curiosity won out, and we soon reentered our vision, learning to become comfortable observing this observer~ We discovered we could play with this eye just as we had with the wheel, in terms of movement and color~ We had a new toy! And just as with the wheel, this mind play continues to this day~ We have also found that, in addition to the geometrical shapes, we can form them into structures that may have a *liquid* type of field, while others may take on something like unto a *foliage* appearance~ Still others manifest as a pulsating energetic medium~

Somewhere along the way, we tried *driving* our *observer* forward into and through the pupil of the eye~ It was not unlike entering a black hole in terms of venturing into the unknown~ However, never having had any kind of truly negative experience other than the brightness events and the initial discovery of the *eye*, we were confident that it was not only safe, but intended as the next logical step in our journey~

What happened next was the most Intellergy-altering experience of our early Thojourns, because when we passed through the pupil, we opened into what can only be described as *deep space*~

(We will begin this particular description in present tense at this time, as this is our current state of meditation~)

The vision is like the view from the bridge of the Star Trek Enterprise, only with many, many times the stars~ We would also add that our stars seem infinitely distant and tiny, yet we are present, immersed, and surrounded at the same time~ There are no apparent large stars or planets; rather we experience ourselves as a *misty star field~* Although we have found we can play with the motion, the star field has never responded to color intention~ We are never inclined to change anything now, even motion, because it is so simple and beautiful~ The singular, most defining element for us in this space is that we are able to be more and more immersed in the space and *become it~*

There is no I, me, my, mine~

There is no body, no name, no identity, no identification, no object, no motion, no relationship — only a sense of what we have come to know as an amazingly full *beingness~*

The frequency and longevity of this experience has enhanced with time, and there are occasions when we are very reluctant to leave this space~ We can envision a time when we will not~

~~~*~~~

At the highest level of Satori from which people return, the point of consciousness becomes a surface or a solid, which extends throughout the whole known universe~ This used to be called fusion with the Universal Mind or God~ In more modern terms, you have done a mathematical transformation in which your center of consciousness has ceased to be a traveling point and has become a surface or solid of consciousness~ It was in this state that I experienced "myself" as melded and intertwined with billions of other beings in a thin sheet of consciousness that was distributed around the galaxy; a "membrane."

John C. Lilly
Tanks for the Memories: Floatation Tank Talks (1995)

~~~*~~~

We were also fortunate in childhood to manifest with the emergence of the graphic novel from DC, Marvel, and the like~ Characters such as Superman, the Flash, the Hulk, Green Lantern, the Fantastic Four, Dr. Strange, X-Men, and others filled our imagination~ We dreamed of life on Earth-Two with Superman's archenemy, Mxyzptlk, the multidimensional mysticism of Dr. Strange, the time travel of the Flash, or the metamorphosis of Bruce Banner into the Hulk after exposure to gamma radiation~ We began to be schooled in the language and basic principles of Quantum Mechanics~ Practically learning to read through these comics, which we read over and over, we began to join in mind journeys concerning the speed of light, multiple universes and dimensions, particle accelerators, enhanced human transformation in intelligence, power, and so on~

On a fundamental level, we identified with these characters and concepts with absolutely no understanding of them on a scientific basis, while we were subconsciously building a vocabulary of quantum terms that made the assimilation of quantum language much more natural later on~

Between the ages of sixteen and seventeen years, we had two near-death experiences~ The first time, we were buried under several tons of road-base gravel material in a railroad car~ The second time, we fell (more like bounced and rolled) from a height of seventy-five feet off Comanche Bluff just northeast of Austin, Texas~ On each occasion, we had to be resuscitated, but we recovered with no lasting effects~ We will not bore the reader with details of the events, because the Intellergetic value of the experiences is what is relevant for us here~ We have to say there was no proverbial *white light* or luminescent *tunnel* associated with these occurrences~ For us, it was more of an extra-dimensional awareness, nothing like our earlier out-of-body space~ This was more like a vivid feeling of self-awareness with no body, no pain, and no anxiety~ As they say, *the accidents happened so fast* that we have no real memory of any initial shock or fear~ The transition seemed almost immediate, as was the return to our physical bodies~ Awakening in a damaged corporeal form had its obvious challenges, but the impressions left by the experiences were life-altering~ Up to this point in our manifestation, death was *out there*, something for which we had no personal reality reference~ Now our transition was a private and intimate truth that forever altered our perception of reality as a manifested being~

In the time following, our Energy Signature began to lose density and became less reactive to energies and forces of conflicting and intrusive natures~ Many humans might describe the Intellergy as *Life is a lot sweeter* or *I don't sweat the small stuff so much anymore*~ We have grown into more forgiving Intellergetic responses to the so-called *daily ups and downs*~ We have found that the more we stay present in the moment, breathe deeply, and attempt to maintain an energy of playfulness at the forefront of our consciousness, the more likely we are to recognize the great opportunities, avoid the grand errors, and basically have the most fun~

Looking back over our path, scattered with so many wonderful metaphysical events, it has become simpler to understand that our Intellergy has been on similar travels through other manifestations for a very, very long time~ We are convinced, for ourselves if for no one else, that all our challenges and experiences with these more expanded realities and perceptions were placed and continue to be placed along our journey to facilitate and accelerate the present evolution of our Intellergy~ They have made it possible for us to manifest at the time of the Great Shift, to communicate more clearly and hopefully become a greater part of the communication *bridge* currently under construction between physical reality and the higher perspective frequencies of expanded intelligence and energy~

Sometime in our early twenties, we spent time with a young woman whose mother had just returned from a spiritual quest to India~ Upon her return, she notified her family that she was renouncing them and all her material possessions~ She had become a Buddhist nun~

At the time, we didn't even know someone could do that!

During our time together, this young woman gave us two books: *The Autobiography of a Yogi*, reflecting the life of Paramhansa Yogananda, and Robert A. Heinlein's *Stranger in a Strange Land*. In these books we found voices and imagery speaking of the higher human potential that had been circling our consciousness for as long as we could remember~ Our path had been set and our quest was the knowing of our true nature, whatever that would prove to be~

This track later led us to *The Dancing Wu Li Masters* by Gary Zukav and Fritjof Capra's *Tao of Physics*. These first encounters with the actual language of Quantum Mechanics and theory came after years of delving into mostly Eastern Philosophy~ The elegant ways these authors blended the similarities between the two completely different disciplines formed understandings in us that resonated very deeply~ The ancient mystics and yogis thousands of years ago had been able to express the most fundamental properties of the Multiverse, reflecting the same discoveries appearing in millennium mathematics and particle accelerators~ These beings had none of these tools, only the perceptions realized from lifetimes of meditation and contemplation in the remotest of caves and on the highest of mountaintops~ A philosophical *bridge* had been discovered that has proven to have great implications for us and remains the fundamental inspiration for this writing~

None of this effort could have been possible without the lives of Gautama, G. I. Gurdjieff, Shankara, Timothy Leary, Ram Dass, Jimi Hendrix, Helena P. Blavatsky, Laozi, Erwin Shrodinger, George Martin, Kim Snider, Albert Einstein, Aurobindo, Nassim Haramein, The Moody Blues, Yogananda, Rudolf Steiner, Wilson Pickett, P. D. Ouspensky, Crazy Horse, J. R. R. Tolkien, Victor Borge, Max Planck, Roy Rogers, MWUSO, Immanuel Kant, Ray Charles, Robert Anton Wilson, William Arntz, Betsy Chasse, Ray Kurzweil, R. Buckminster Fuller, Franklin Merrell-Wolff, Jim Morrison, Neil deGrasse Tyson, Sun Tzu, Kelci Auliya, Keith and Maura Leon, Bobby Bland, Thaddeus Golas, Robert Pirsig, Joe Tex, Greg Betts, Nelson Mandela, Louis Armstrong, Candace Pert, Fred Alan Wolf, G, Smokey Robinson, Don Shepherd, Veo Merida, Goldie Goldberg, Little Richard, Dr. Spencer Brown, Jane Roberts, Erin Nicole, Walter Y. Evans-Wentz, Hermann Hesse, Ayn Rand, Robert Heinlein, Jimmy Durante, Carl Sagan, Leon Russell, Rod Serling, Tony Von, Edgar Cayce, Goldie, Freddie King, Gene Rodenberry, Thich Nhat Hanh, Red Skelton, Lewis Carroll, John, Paul, George, Ringo, and countless others along the way~

Throughout the above process of *unfolding*, we felt inside the faint but growing ringing of fundamental truth~ Over time, the ringing has become an insistent clanging of alarm~ To find ourselves in the realization that we had literally been living in a world enslaved by our beliefs, in a reality that did not exist, became unacceptable~ This understanding was an intimate *awakening* for us, and at once we knew what the great sages and teachers had been speaking of for thousands of years~ In a very short time, our hobby of curiosity grew into a passionate obsession~

> *Awakening is possible only for those who seek it and want it, for those who are ready to struggle with themselves and work on themselves for a very long time and very persistently in order to attain it~*
> — G. I. Gurdjieff

> *Concerning matter, we have all been wrong~ What we have called matter is energy, whose vibration has been so lowered as to be perceptible to the senses~ There is no matter~*
> — Albert Einstein

> *If you want to find the secrets of the universe, think in terms of energy, frequency, and vibration~*
> — Nikola Tesla

So . . .

With these statements in mind, here is how our thought process began to unfold:

If Quantum Mechanics is correct and there is no matter, then that would mean that everything in the Multiverse is simply energy, including the entire physical universe, from galaxies to life forms~

> *The good thing about science is that it's true whether or*
> *not you believe in it~*
> — Neil deGrasse Tyson

> *Just because you don't see it*
> *doesn't mean it isn't there~*
> — MWUSO

For us to perceive things like our body, our house, our town, or our world as energy alone, that energy would have to be organized as information; it would have to be intelligent~

If our perception of this energy is so low in frequency that it appears as solid matter, then would it be possible for us to raise our frequency in ways to perceive the true nature of our reality and our place in it as the intelligent energy that it is?

What does this realization mean for everything we always thought and believed to be true?

Can we just continue to live, work, play, love, or think in the same old ways, knowing it is all based on false understanding and perspective?

What would our potential be if we could learn to live from what we really are: beings of intelligence and energy unencumbered by physical bodies?

If the entire universe is intelligent energy, then what would be our potential if we could tap into and blend with that One Mind?

> *There is a point when all must be lost*
> *In order to find what might be gained~*
> — MWUSO

Earlier in our little story, Horace risks everything he had ever believed about his body, his home, his friends and, in the end, his entire world for a *calling* from something greater than himself and his beliefs~ Once he had experienced this larger, brighter reality, he was no longer able to return to his old beliefs and exist in the same ways~

This will be our challenge to you and all humans in this quantum world~

It is our own intimate and personal truth that we are not just fragile, limited physical bodies with mind/brains driven by the desires and expectations of derivative, formulaic, and predictable thoughts and beliefs~ We are fluid, creative beings of unlimited energy and intelligence infused into a One Mind that has emerged from an unknowable and indescribable Source~

This is also your true nature~

Isn't it about time you got to know yourself for what you really are?

As you continue to cross our threshold, please be so kind as to remove all of your existing belief systems, along with any derivative, formulaic, or preconceived thought streams, and drop them off here at the entrance~ These might include certainties regarding your physical body and its sensory perceptions, nationalistic or political loyalties, religious convictions, professional endeavors, or personal relationships~ You may pick them up in their original condition as you depart if you so desire~

There is no cost of admission other than the price of this simulation, although it is our sincere hope this was passed along to you randomly in well-worn and dog-eared condition by a stranger on a beach~

~~~*~~~

In the province of the mind, what one believes to be true is true or becomes true, within certain limits to be found experientially or experimentally~ These limits are further beliefs to be transcended~ In the mind, there are no limits . . . . In the province of connected minds, what the network believes to be true either is true or becomes true within certain limits to be found experientially or experimentally~ These limits are further beliefs to be transcended~ In the network's mind, there are no limits~

John C. Lilly
The Human Biocomputer (1974)

~~~*~~~

You are most likely already beginning to sense the stimulation of certain harmonics and resonances within your Energy Signature as a result of following these simulated pages of Intellergy~ We encourage you to allow these timbres to evolve melodically with as little prejudgment or interference as possible, so as to permit the greatest tuneful synergy~

This book, which we will call a *simulation*, is a playful manifestation in the One Mind, recognizing that all things — including energy, consciousness, information, imagery, language, objects, thought, action, institution, religion, politics, dualism and the rest — are illusory, emergent aspects of an unknowable, indefinable Source beyond all human understanding~

Please do not seek any sort of traditional order or context~ There will be no definite course of study or action~ New terminology will be introduced without definition so as to be discovered at a later time, although you may refer to the Expanded Glossary at the back if you wish~ The reader has probably noticed that we have made the literary choice to use a *tilde* instead of a period for punctuating the ends of our sentences~ One of our early readers reacted to this grammatical choice as an *uncomfortable tic*~ It has become a convenient coincidence that this little piece of punctuation art is a subtle reminder that each sentence, as well as the entire work, is a collective progression of collapsed wave functions that have become fixed in time and space, vastly restricting further Intellergetic evolution~ There are many outside sources explaining the *wave function* and the *probability cloud*, but its relationship to our truth should emerge as you read~

It has been found that those who are reluctant or unable to leave their belief systems and expectations behind, which are based on other writings or experiences, often become dissociative, going through a blurred sense of personal identity in extended contact with Blindsight perspectives~

This is particularly true of what is known as the *Intellectual Mind*, which is made up of very elaborate mental constructs, preconceptions, expectations, and confident methodology~ We especially reach out to these magnificent intellects, requesting that the following Intellergy be allowed to unfold gradually, without restriction, blending into a flow~

Celebrate the alternatives~

Enjoy the ride~

The medium is the message~
— Marshall McLuhan

Blindsight is a medium and a process for personal discovery that is infinite in nature~ This simulation is random in design~ At the point one becomes comfortable with the new terminology introduced, the stream may be stepped into at any point; chapters may be read in forward or backward order and hopefully still provide relevant content~ Many of the writings that have been integral to our own unfolding awareness were assembled in this way~

There will be no finite belief system or path of action offered, other than that limited by one's own intention~ The hope is to ignite a passion for discovery in millennium mind/brains of their true nature—as intelligent energy beings entangled in an infinite Multiverse~ The ultimate goal is to join in the world-wide effort to raise the cumulative Intellergy of the planet, generating critical mass for a fundamental shift in awareness~ Any existing problems or issues currently facing humanity will certainly find quicker and simpler solutions when met with higher energy and intelligence~

We are self-aware, intelligent energy~
This is our truth~
It is our free choice to live from this truth~
Anyone existing as a physical brain body
has the same truth; the same choice~
— MWUSO

This work is ultimately a virtual reality made up of energy and intelligence~ As such, most of the relationships, events, imagery, and expression will be conveyed in a language of Intellergy reflecting the true nature of the perceivable Multiverse~ Readers may find this technique of expression uncomfortable or even pretentious at first if certain belief systems or thought streams were not completely left behind earlier~ But it is our hope that as we progress, more familiarity and understanding will follow~

Please remember, after all, that you are, in quantum terms, simply a being of energy and intelligence, so this is really the Intellergetic language of you~

Within a few chapters, we are hopeful you will begin to find increased fluidity of thought and augmented expansion of awareness~

We do not come from any kind of scientific, quantum, or philosophical background and have no formal training or classwork in any of these fields~ As a result, most of this simulation could be categorized as a work of fiction springing from meanderings of the well-intentioned mind/brain of a common Energy Signature~ We write more from a position of mystery than mastery over these concepts~ We are struggling as much with our evolutionary path as everyone else~ It is simply that we seem to have developed an aptitude for communicating ideas of this nature in a different sort of manner, and we hope that our streams of thought can help others illuminate their own possibilities along the way~ All mistakes are ours in this particular Multiverse, allowing that in certain other parallel Multiverses, these errors may resonate with fundamental truth~

Often throughout the simulation, we will request that the reader stop and pause to think more deeply about a concept~ We may ask you to perform a simple task, such as taking a deep breath or participating in a mind experiment~

Please pause and take your time in these passages~ It is common for an experienced reader to build momentum in reading and pass over opportunities such as these~ We want to remind you that the journey here is the most important aspect, so please relax and enjoy~

> *Don't push the river*
> *It flows by itself~*
> — Barry Stevens

In late-night conversations over bottles of wine, many have suggested we undertake such a project~ It would appear that time has come~

~~~*~~~

The most evolved, Intellergetic human in our energy sphere is a healer who works from *can 'till can't* each day, administering to the poor, indigent, elderly, and homeless, never accepting any compensation~ This being will never read this book~ We can just hear them saying, "This is so-o-o complicated; your energy must be simple and pure, intimate and known." They would also remind us to find stillness in our self, even when there is none in the world around us~

Another of our extremely evolved friends would wholeheartedly agree, but at the same time they would offer that a *connection* is being called for at this unique point in human evolution that is being called the Great Shift~ They and their collection of millennium Light Bearers are part of shaping a movement globally linking and uniting beings of higher Intellergetic frequency awareness to create conduits and portals for sentient beings caught and confused within the accelerating densities and tensions of the millennial context~ Their offerings and works are examples and pathways for an intentive seeker to navigate blustery Qen4 turbulence, while learning to resonate with the Qen5 harmonics in the eye of the storm~ Blindsight is a part of this undertaking~

From a very early age, we have never been overly interested in the who, what, when, where, and how questions~ However, the *why* of everything was always an embedded curiosity that later developed into a borderline obsession~ It has now become more of a comfortable passion~

Why do we exist?
Why do we know we exist?
Why do we know that we know?

In order to best convey ideas that will, at first, appear disturbing and possibly confrontational, we have chosen to create some new language for this Quantum World in which we live while adapting some of the old~

For example, we choose to see ourselves, all objects, language, thoughts, and phenomena as intelligent energy~ The word we have crafted to express this is *Intellergy*~

We perceive ourselves, other humans, and all of reality as condensed accumulations of Intellergy~ Our words to describe these condensations are *Energy Signature*~ *Signature* is chosen because humans insist on naming, measuring, and quantifying every perception in their lives~ From a Quantum standpoint, the only thing that is not changing at near-light speed in your life every moment is your *name*~ We exist in this universe, spending more than 99 percent of our perceived lives as the unlimited potential that is possibly Dark Matter, Dark Energy, Neutrino Oscillation Anomalies, or others~ If we were able to observe ourselves from a higher frequency perspective or with vision capable of near-speed-of-light resolution, we would appear as merely a temporal *flickering* rather than something solid~

(Search out the outstanding young physicist, Nassim Haramein, with his beautiful, humorous, and simple explanations of this and many other quanta fringe concepts~)

Your name, personality, and identity perception becomes a sort of *gravitational center* for attracting Intellergetic ideas, beliefs, and perceptions that continue to draw in more of the same Intellergy in an ongoing cyclical process, establishing and reinforcing your identity and personality~

~~~*~~~

Please, for the sake of accelerating your participation in the upcoming simulations, close your eyes at this time for one or two minutes and try to envision your body as its true nature, of pure energy composed of intelligent information~

Attempt to picture your Intellergetic body as a glowing, swirling, pulsating entity in constant flux~

This body should be spherical, the same as all other perceived stars and planets in this Multiverse~ The Great Attractor at the center of this body is your name, your identity, your personality, your I~Me~My~Mine Intellergy~

Every single bodily function, thought, word, or action is invoking reactions in the energy of the Signature, harmonically resonating with accumulated experience, belief, and expectation~ These reactions continue to attract like densities, reinforcing the systems of belief and perception~

Within this being, imagine areas of brightness, with other regions of discoloration or darkened appearance~

These brighter regions would reflect self-aware, expanded intelligence with extremely high frequency and fluid potential for forming your reality~

Darker, accumulated densities would reveal accretions of reactive, derivative information of collapsed wave functions that continually cycle in upon themselves, attracting more of the same compacted and limited denseness manifesting as reinforced habit and predictability~

*Once you can actually **see** yourself as this being that is your true nature, instead of a bag of skin and bones, you will be able to begin imagining ways to understand, observe, and create from your Intellergy~ This will be the beginning of the healing process from the contamination of the Human Disease~ You can start dissipating the densities that limit your evolution and transition into higher-frequency perceptions~*

~~~*~~~

From this point forward, we will refer to our *life* as our **Manifestation**~ This word reflects our true nature as beings of energy and intelligence~ Its expression becomes a way to begin thinking of ourselves as more than just a meat suit~

In that same spirit, we will begin to refer to this book as a **Simulation**, recognizing that it is also a manifestation of energy and intelligence~

Quantum Mechanics compels that there exists infinite universes and multiple dimensions within them, so we will refer to our current environment as the **Multiverse**~

Blindsight is the name we give this discussion of consciousness and energy in relation to Quantum Mechanical principles and ancient contemplative philosophies~

Blindsight compels that something exists outside the Multiverse, and that it is expanding into this~ Blindsight also necessitates that this same *something* existed before the Big Bang~ This *something* is conceived as an infinite *field* of intelligence and energy that subsists only as potential and possibility with no form, movement, or object~ This field we will call **Qen**~

From the Expanded Glossary, here is our definition of **Qen**:

In Blindsight, Qen is ultimately indefinable, and yet a term must be conjured here for the expansion purposes of this simulation~ Qen is the infinite field, pool, continuum, or medium of potential intelligent energy from which all manifestation emerges~ Qen is the undisturbed *palette* from which the indescribable *Source* mixes with intention the many *colors* of creation~ Qen may Intellergetically manifest as a fully expanded spatial aspect when observed as such, but Qen may also be perceived as particulate with characteristics that simulate mass~

Qen, for this Multiverse, exists as a Qentinuum, within which the Multiverse emerges~ This is to include all manifestation, Intellergy, and Spacetime~ The closest analogy is the Quantum *dimension*; however, all dimensions evolve within Qen and emerge from Qen~

Qen, in human perception, assumes an infinite spectrum of frequency expansion when spanned across this Multiverse~ This spectrum has responses reflecting *strata* of harmonics and resonances like an FM radio station at 101.5 MHz that samples from the FM radio band between 88 and 108 MHz~ These strata, although completely entangled with the full spectrum, resonate with bands of frequencies that take on multiversal properties loosely individual to the respective sample of the spectrum~ Much like the radio station, these bands assume densities for resonant perception that become Intellergetic in structure~ Qen, in this context, may be assigned harmonics, such as Qen4, which would reflect our physical perceptions or Qen5 sensitivities that would reveal perception from a much higher frequency~ These finer resonances afford an Energy Signature the potential for immersion into the Source and/or experience for transition movement across Qen strata through Thojourn, meditation, hallucinogen, or some other form of pro-active, mindful Intention~

Qen remains unmoved and formless until it is infused with *Intention~*

> *Intention is one of the most powerful forces there is~*
> *What you mean when you do a thing*
> *will determine the outcome~*
> *The law creates the world~*
> — Brenna Yovanov

> *Our intention creates our reality~*
> — Wayne Dyer

> *A good intention clothes itself with sudden power~*
> *When a god wishes to ride, any chip or pebble will bud*
> *and shoot out winged feet, and serve him for a horse~*
> — Ralph Waldo Emerson

One way to understand this phenomenon of Intention is to take the example of a video game, such as Star Wars, which has developed into an online culture with millions of players worldwide~ This *cyber world*, consisting of multiple civilizations, solar systems, environments, and personalities, exists only as unformed potential on personal computers and servers~ All battles, avatars, locations, movements, or choices are only possibilities until they are activated by the **Intention** of the player~ The *reality* of the game environment is simply potential energy and intelligence residing on hard drives, cloud servers, and random access memory, awaiting Intention from the keyboard or controller~

Consider for just a moment that your mind/brain works in exactly the same manner~ For example, please think of the person most dear to you at this time~ Understand that there is no *hard copy* saved in your brain for their appearance, your experiences with them, or your feelings for them~ All of these perceptions exist as potential intelligent energy waiting to be retrieved from your *data banks* by your intention, which you just inputted into your mind/brain~ These *perception programs* will continue to *unfold* in present time, either reactively as you allow the stream of consciousness to flow or proactively if you so choose in a mindful way to conjure a memory of a specific experience or feeling~

Is it really such a stretch to relate to our own Multiverse in the same manner? To see it as unlimited probability and potential, pending our next thought, action, deed, or inspiration to create the next piece of our reality?

The most integral tool we will offer in the coming simulation will take the form of intentive creative visualizations and thought journeys~ We will refer to these as **Thojourns** to convey a millennial intention~

In Quantum Mechanics, there is an essential piece of every calculation, experiment, and theory that must be included and cannot be ignored: the role of the **Observer**, even though this Observer has never been observed~

The observer is the observed~
— Jiddu Krishnamurti

. . . separation of the observer from the phenomenon to be observed is no longer possible~
— Werner Heisenberg

You are an aperture through which the Universe is looking at and exploring itself~
— Alan Watts

First recognized in the observation of the dual properties of light early in the twentieth century, Max Planck and Albert Einstein proved that light can be perceived as both a particle in nature and a wave form~ How the light appears depends completely on how it is observed or how the experiment to observe it is set up~

One might also refer to the thought experiment *Schrodinger's Cat*, where a live cat is enclosed in a box with a randomly decaying radioactive isotope, which can at any point reach critical mass and kill the cat~ The crux of the experiment is that until an *observer* opens and looks into the box, the cat is theoretically both alive and dead~

Over the last century, countless experiments and theories have tried to identify, define, measure, or observe the Observer, to no avail~ Every attempt ends in the inevitable collapse of the wave function performing a selective outcome~ Each time, the event dissolves into duality~ It is a *particle* or it is a *wave*; the cat is *alive* or it is *dead*~

Albert Einstein was even heard to ponder whether the moon was actually there when we are not looking at it~

All the answers for this question of an Observer seem to always return to a conversation about a Supreme Being with a name like God, Atman, Allah, Brahman or others~ In Exodus 3:14, Moses asked God for his name and the response he received was, *I am that I am*~ Moses needed some kind of word to describe the indescribable~ *Jaab Sahib*, the morning prayer of the Sikhs, composed by Guru Gobind Singh, contains 950 different names for God~

In Blindsight, the use of a word or name for the so-called *Supreme Being* is, if not blasphemous, at the very least absurd~ For our purposes in this narrative, we will very simply use only _____~ If, for the time being, the reader would choose to insert some sort of personal identifying reference or word, they of course certainly may; however, it is our hope and goal that said reader will leave our discourse without the need for such naming of the indescribable~

I AM the nature of pure consciousness~

I AM the same to beings, one alone;

I AM the highest Brahman, which, like the sky, is all pervading, imperishable, auspicious, uninterrupted, undivided, and devoid of action~

I do not belong to anything since I AM free from attachment~

I AM the highest Brahman . . . ever-shining, unborn, one alone, imperishable, stainless, all-pervading, and nondual~

That I AM, and I AM forever released~
— Shankara
The Upadesasahasri

Shankara was not speaking of his personal self as some sort of god in a physical body, but rather in the *Impersonal I* that is _____, exclusive of any form, object or action~

In our own meditations, the *Observer* is _____~

_____ is the source, the creator, the realized, sempiternal, multiversal beingness~

Our one goal is simple immersion into inexpressible beingness~

Any implication that we would offer up some sort of supreme being existing outside of our own oneness with _____ would be a complete misunderstanding~ _____ is the unknowable source of all manifestation and is infused within, throughout, and exclusive of all energy and consciousness~

_____ is the self-aware source of creation, observation, and intention that exists before, after, and permeates all Multiverses and dimensions~ The Blindsight position maintains that all names imply a dualistic relationship between the self and a supreme being~ This will never be our intention~ Rather, we will attempt to establish that all duality (me~you, here~there, now~then, this~that, good~evil) is illusory~

When one is fully absorbed in _____ while still in human form, this state is referred to as **Signome**, and it is the ultimate bridge between Qen4 and Qen5 Intellergy~ Signome reflects a state of awareness that is most similar to the Nirvana, Satori, or Samadhi of the ancient and present-day mystics~

The very endeavor we make to define or discuss _____ here is an exercise in absurd futility~ Nevertheless, oneness with _____ is the ultimate culmination of our manifestation~

Within a few chapters, we are hopeful you will begin to find increased fluidity of thought and augmented expansion of awareness~ Should you encounter any of the unfamiliar words conjured for this simulation, please refer to the Expanded Glossary at the end or simply allow their meanings to unfold in context~

Onward through the fog~
— Oat Willy

NOTES

Daonna Galar by Kelci Auliya

III

The Human Disease

What will your future be?
What kind of life will you have?
Will you be successful?
Will you be happy most of the time?
Will you find great love?
Would you want to create the kind of life you want?

These types of leading questions seem to drive the bulk of content in the self-help materials we have all seen over the last half century or so~ Millions of humans, including ourselves, have found meaning and paths of action to better our lives, achieve our goals, and deepen our relationships~

At this time, we would like to *fine-tune* this process of self-evolution~ Some of the questions might be:

What will the next five seconds of your life be like?
Will your next perception be true and clear or will it be clouded by expectations and preconceptions?
Will your next act in life be mindful, deliberate, and intentive or will it be reactive, predictable, and unconscious?
It should be fairly simple to see that our entire future hinges on our very next thought, observation, or action~
How can we become more present in our lives and take charge of the next five seconds and the next?

There are many misconceptions about our true nature as beings of intelligent energy~ Our lack of understanding of our true nature makes up the symptoms of what we will call the **Human Disease~**

We have it~
You have it~
All of us have it~

~~~*~~~

The young insurance sales representative had an appointment with an elderly woman to show her options for supplementing her government health benefits~ The woman had been prequalified by an associate in his office through a phone interview, and the sales associate had been assured the woman had no pre-existing conditions, such as illnesses or syndromes that would prevent her coverage~

He arrived at the appointment, was greeted by the woman, and they sat down to begin~ The young man started by getting her basic information when the woman's daughter arrived and asked what was going on~ He explained he was helping her mother with her insurance~ The woman smiled and apologized for his trip because, she said, "I'm afraid that would be unlikely, because my mother has Alzheimer's disease, and she is technically uninsurable."

The young man said that was not possible, as she had been prequalified and that she had told his associate that she had nothing of this kind~

The daughter replied, "Yes, that is probably true because she can never remember she has the disease."

~~~*~~~

Can you even imagine having a life-threatening or life-altering disease and not knowing that you have it?

Nevertheless, this is exactly what we all suffer from~

The Human Disease is not a bacterial infection treatable with a course of antibiotics~

It is not a virus that we can allow to run its course~

It does not just go away with plenty of liquids and rest~

The Human Disease is an entrenched, debilitating syndrome reinforced by collaborative belief systems in humans over tens of thousands of years~

~~~*~~~

These are beliefs such as:

This is a physical world made up of mass and matter~

We walk around in fleshly bodies of skin, blood, and bone~

Each human has a singular, independent identity, and personality in relation to all other humans~

An individual human has no existence prior to physical birth~

Most waking human choices, actions, words, and thoughts are mindful, conscious, and proactive~

This universe, which we perceive is the only one in existence, is made up of galaxies, stars, planets, black holes, and the like~

Nothing existed before this universe and it is expanding into an empty void~

Intelligence does not occur anywhere but in life forms~

~~~*~~~

There are four basic causes for this Human Disease and these are **Language, Attachment, Addiction to Tension/ Release,** and **Reactive Nature:**

Language ~

In the beginning was the Word . . .
— John 1:1

Just as biblical tradition implies that the word of God initiated the creation of the universe, we as humans allow words and their meanings to create our Multiverse of understanding, kinship, and expectations with our environment, our relationships, our self-image, and ultimately reality itself~

In quantum mechanical theory, all combinations of perceived mass, energy, and information exist as infinite, undisturbed potential in a state of superposition with several eigenstates~ An *eigenstate* is the condition of an object *(remember for the sake of this discussion that a thought is also an eigenstate)* that can be measured in terms of its position and momentum~ Prior to observation or measurement, objects exist as infinite possibility in what is called a *probability cloud~* However, once the phenomenon is observed or measured, what is known as its *wave function* collapses, taking on defined reality and density~ A wave function can be recognized as either a particle with position and mass or as a wave with momentum, but never as both~

Within the probability cloud, earlier we wrote about Schrodinger's Cat, but for now please understand that the moment Intellergy is observed, identified, measured, or defined by word(s), infinite potential has vanished~ The wave function has buckled~

There is a singular moment of unlimited possibility with the expanded potential of a mindful act of Intention in a present *NOW* moment~ This we will call **XPotential**~

Here follow some examples of how language impacts and limits human reality~

If one asked most humans what one word would be used to describe the event that happens at end of the day when one sees the sun disappearing over the horizon, most would say it was a *sunset*~ In actuality, the sun is not setting at all~ It is the earth that is rotating away from it, giving the illusion of the sun going down~ Billions of us are raised from children with this misconceived reality as a result of language without ever questioning it~ How different would the experience become if we sat on a mountaintop feeling the earth under us rolling away from the sun, instead of having the perception that we are stationary as the sun drops below the horizon? Shouldn't we be giving our children this experience?

A person sitting in a chair might remark that they are not moving at all, when in reality the earth is spinning at close to one thousand miles per hour and whirling around the sun at around sixty-six thousand miles per hour~ Our solar system is simultaneously traveling at more than five hundred thousand miles per hour in its journey around the Milky Way galaxy, drifting in and out of the spiraling rings~ We also slide up and down from the denser, more *equatorial* regions of the galaxy to the finer regions at other latitudes~ The Milky Way is also traveling toward its inevitable collision with Andromeda at about three hundred thousand miles per hour~ Just to keep it interesting, there is something called the *Great Attractor* out there that is drawing us toward it at 392 miles per second~ Cosmologists theorize it to be a massive object tens of thousands of times more immense than even our own substantial Milky Way galaxy~ The Great Attractor creates gravitational anomalies in deep intergalactic space~ Much more information can be found on this phenomenon if so desired~

For this person sitting comfortably in their easy chair, their reality is that they are living, working, and relating from that motionless perspective~ Their only sensations of motion are generated from their bodily movements, walking or riding~ This is their reality~ How much richer would existence be if we lived from the perspective of an energy being on a big, blue Intellergetic ball rolling at thousands of miles per hour through endless space? Imagine the difference in the evolutionary intelligence of a child who learns to live and dream from this perspective~

> *Hey there, little Earthling*
> *Just along here for the ride*
> *Take a look around you now*
> *There's no place left to hide*
> — The Tortugans

If one were to observe another human tossing a ball into the air and they were asked to remark which direction the ball traveled, they would most likely say the ball went *up then down*~ Imagine if someone on the other side of the planet were able to observe the same phenomenon, they might say the ball went *down then up*~ How would this appear from the moon, left to right?

The reality is that the ball is moving *in and out* from a gravitational center, and yet humans again live with this limited and false reality~ How much more engaging and interesting could a child's play become when empowered with this realization of their play's true nature?

We see with our eyes~
We hear with our ears~
We smell with our noses~
We feel and touch with our skin~
We taste with our tongues~

In actuality, none of these statements are true, but because of their common use and accepted versions of reality, they collectively contribute to ever-expanding misconceptions and illusions about how humans view and experience their physical and Intellergetic forms~ In fact, all these organs of perception are merely sensors attracting, receiving, processing, and transmitting environmental information for the mind/brain to further process, store, and transmit~ How much more interesting would these sensory perceptions become if viewed as input from densities of energy and intelligence entangled with One Mind's Intention?

We now know our solar system extends at least one hundred times the distance of Pluto from the sun~ The Oort Cloud is a gathering of dormant comets still under the influence of the gravity, cosmic and electromagnetic radiation of our sun~

As Intellergetic Beings, we must strive to expand our own *sensory array,* understanding that our ability to receive, store, process, and transmit Multiversal information is infinite and not that of meat sack limitation~ Our physical senses appear Intellergetically as transitory blips emerging from a vast, entangled Energy Signature~ It is only our attachment to denser harmonic frequency perceptions that affords their readings any validity at all~

Humans are capable of wonderful stereoscopic vision, and it is the strongest sense, and yet an eagle has 3.6 times the acuity~ Certain insects have a compound fisheye view that allow them to see objects far and near simultaneously~

Human hearing range is up to about 20,000 hertz, while the Bottlenose Dolphin's range is up to 105,000 hertz~ Even in our sack of meat, we have developed an arrogance about our perceptions of our environment~ We feed this superiority that we perceive with the language of expectation and duality~

One might argue that these little mind games are petty or irrelevant and have nothing to do with being able to live and survive~

We would counter that these are merely several of a myriad of symptoms of the Human Disease spread throughout our Intellergy Forming language~ These symptoms thread all through the fabric of a global human collective reality and continue to shape and limit belief systems, understanding, and our ability to live from our true nature as Intellergetic beings with infinite XPotential~

Ask ten people on the street what the word *love* means to them and you will get ten different answers~ Ask them to describe *sweet* or *hot* or *politics* and one will quickly find that each person's responses convey densities within their Energy Signatures relative to the expectations of their Intellergetic interactions with the words. Yet, many of us venture forth through life with borderline arrogance in our understanding, as well as our ability to relate ideas or concepts that are extremely limited and fundamentally misunderstood by everyone with whom we communicate and who have their own relative density-resonant expectations, and interpretations~

All language is as the Zen *finger pointing at the moon*~ Language, in the end, is merely a mechanism humans developed for survival, community, defense, and every other aspect of the interpretation of reality~ It is not and never will be any true actuality~

There is a field of study called Neurolinguistics that draws its theories and methods from many disciplines, but one of its core approaches explores how language images assimilated in the brain create our reality, perception, and expectation~

~~~*~~~

John was talking over the fence one afternoon to his neighbor, Robert, about the explosion of the Grackle population in their area and the problems they were creating~ He went on about how they multiply prolifically, will eat almost anything, and have no natural predators~ They spread diseases and can be disastrous to certain citrus crops~ These are only some of the problems they bring with them, and he remarked that there were literally hundreds of thousands of them in the region~

Robert replied that he had no idea that undocumented aliens could cause such difficulties~

For some reason, Robert had understood that Grackles were something completely different~

"No, I mean blackbirds!" John exclaimed, as he finally realized Robert had no idea about what they were talking~

~~~*~~~

Robert had simply never heard of Grackles, and as a result the communication was defective~ How many times do these types of exchanges take place every day, not only in our personal manifestations, but also on the highest levels of human interaction and communication?

In sales training books and seminars, associates are encouraged to listen to how customers and potential buyers use language so they can use the same words and inflections back to them in their presentations, demonstrations, and closing techniques~ Subliminally, a customer hearing familiar turns of phrase and inflection will have their robotic densities stimulated in accustomed frequencies, increasing the susceptibility for purchasing~ Sales associates also learn to *mirror* the customer in body language for the same reasons~ If a customer puts their hand in their pocket, the associate will do the same~ The associates find that if they nod and smile when emphasizing certain features and benefits of their respective product or service, the buyer will very often do the same, rousing other energy centers and leading to buying proneness~

One of the saddest symptoms of the Human Disease is the human fixation that language is real~ How many Porsches, Time Shares and Jimmy Choo boots have been bought under the spells of neurolinguistically-savvy sales people? How many misunderstandings, dysfunctional situations, shattered relationships, even wars, have been the result of a mistaken inference drawn from a statement or document?

Ironically, we will have to continue using the paradox of language here to convey higher and deeper Intellergetic meaning~ Hopefully, the exceptions to our process will become clearer~

Another compelling aspect of our existence that language ignores is that **nothing in the universe ever physically touches anything else~**

Quantum Mechanics compels that the atoms that make up all matter are swirling entities of energy and information, transforming in this Multiverse at near the speed of light~ Only the energy and information are exchanged~ Even in the older Newtonian models of the atom that portrayed a solid nucleus composed of protons and neutrons with electrons like billiard balls in orbit around them, any contact between one atom and another would only result in a collision or exchange of electrons, which, of course, could never happen~ There would never be actual contact between so-called objects, fluids, or even gases~

We understand now that only energy and information are traded; this is true of a handshake, an asteroid collision, a ship on the water or a kiss~

How far removed are you from your true infinite, entangled nature if you perceive yourself in physical contact with your world?

How much richer would the experiences of your senses be if you saw yourself entangled with a Multiverse as an Intellergetic being?

As an Energy Signature evolves and grows based on attachments to Qen4 realities involving a physical body, a personality, possessions, time awareness, motion and position, their density continually builds, attracting more and more resonant harmonics compounding the illusory Intellergy, ultimately defining their reality and self-image~ This perspective is the root of I~Me~My~Mine consciousness~ As demonstrated earlier, self-involved density is the origin of duality, tension, fear, and dysfunction~

Attachment ~

The second fundamental factor identifying the Human Disease is the human propensity for *Attachment*~

Detachment is not that you should own nothing, but that
nothing should own you~
— Ali Ibn Ali Talib

Attachment to things drops away by itself when you no
longer seek to find yourself in them~
— Eckhart Tolle

There is actually a psychological model called *Attachment Theory* that explores the dynamics of interpersonal relationships beginning in childhood and expanding into adulthood~ It is fair that attachment Intellergy comes into play at the inception of a manifestation with the connection to a caregiver or the lack thereof~ From the very beginnings of life, we attach ourselves to objects ranging from the breast or bottle to toys, clothing, or blankets~ These connections are usually reinforced by surrounding Energy Signatures whose Intellergy is constructed of their own accumulation of densities stemming from their first realizations of their manifestations~ The cycle soon becomes self-perpetuating, delusional, dysfunctional, addictive, and frighteningly destructive~

One thing alone is certain, that man's slavery grows and
increases~
Man is becoming a willing slave~
He no longer needs chains~
He begins to grow fond of his slavery, to be proud of it~
And this is the most terrible thing that can happen to a man~
— G. I. Gurdjieff

Physical reality is one of the biggest horror movies of all,
and you know how we love horror movies~
— Thaddeus Golas

The root of suffering is attachment~
— Buddha

The fundamental problem with language as a factor of the Human Disease lies in attachment to the power and imagery of words~ Humans are willing to sacrifice, fight, and die over attached beliefs with regard to their interpretations of religious, political, or personal language~ They are also prepared to lie, manipulate, or even kill for the same attachments~

You talkin' to me?
— Robert De Niro, *Taxi Driver*

Now we come to one of the ultimate dilemmas resulting from these first two symptoms of the Human Disease: ***Attachment to Language and the Language of Attachment~*** This quandary is a compound symptom of the Human Disease that must be recognized by each Energy Signature for themselves~ There is an impasse created by this combination that is a Qen4 squirrel cage, from which there is simply no escape without the hard work of self-actualized, *NOW*-moment, intentive observation~

Once our words can be clearly and intimately known as only an energetic, fluid resonance of Intellergy, they begin to lose all their power to influence us, whether positive or negative~ The exceptions are words written, spoken, or even sung with self-actualized, mindful intention~ These timbres often begin to alter, influence, and finally shape physical, emotional, and mental realities with higher frequencies~ These expressions of true nature are able to engage Qen5 Intellergy in Qen4 consciousness~

The process of extricating oneself from the infections of the Human Disease has, for centuries, been referred to as *Awakening~*

We are near awakening when we dream that we dream~
— Novalis

When one realizes they are asleep,
at that moment one is already half-awake~
— P. D. Ouspensky

Those who are awake live in a state of constant
amazement~
— Buddha

The idea of *awakening* implies that one is asleep~ The Human Disease is, if nothing else, a state of sleepwalking~ So attached to the illusory images, objects, and language of Qen4 material existence has humanity become that we walk as robotic slaves in a dream state, oblivious to the wonder of our infinite true nature just a flash of wakefulness away~ We live in constant entanglement with other enabling Energy Signatures for whom the Human Disease is a natural and acceptable state of reality~ One who suggests or acts differently is most often viewed at best eccentric and at worst insane~

The first to awaken are often disgraced
before they are embraced~
Being wakened from a slumber is not always welcomed by
those clinging to the final moments of sleep~
— Jeannine Sanderson

Being considered "crazy" by those who are still victims of
cultural conditioning is a compliment~
— Jason Hairston

How many of us hit the snooze button in the morning when the alarm sounds? How many set alarms very early, actually planning on not getting up when the sounds go off, allowing for one or multiple snoozes? We have one acquaintance who actually places an alarm with a most obnoxious sound in another room, forcing him to physically get up and turn it off so he won't fall back asleep~ It is not easy to awaken from a comfortable slumber, especially if waking up represents an immediate future of work or tension or anxiety~

One's true nature as an energetic being is always intimately present, awaiting our immersion~ The only factor blocking, clouding, or preventing this union is our slumbering attachment or *slavery*, as Gurdjieff called it, to Qen4 language, objects, and imagery~ Unfortunately, once an Energy Signature has achieved a certain level of attachment to the densities of Qen4, a confidence arises that this is a territory to be learned, manipulated, conquered, and eventually understood~

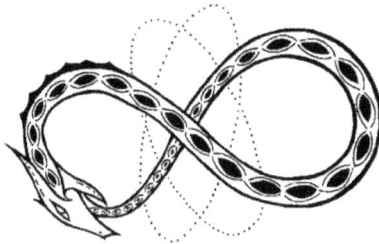

Fosghair Ouroboros by Kelci Auliya

Spacetime is curved in the physical universe, reflecting some quantum mathematics that would have a traveler reaching for the end of the universe eventually returning to his starting point~ Much like the Ouroboros, the ancient symbol of the snake eating its own tail, those oblivious to their affliction with the Human Disease fall into a cyclical existence, ever building up densities of preconception, habit, and self-destructive behavior~ These Intellergies continually chase and consume their own *tails,* recycling derivative, reactive densities~ At the same time, they are continually attracted to those harmonics of energy and information that exist in more contracted or complex manifestations~ It is like hitting the Intellergetic *snooze button* on one's evolution~ This cycle of density attraction is the ever-building pursuit of many disciplines, political platforms, educational traditions, religions, and practices that would proclaim pathways to _____, when in truth they only create more duality and complexity by Intellergy Forming realities that perpetuate their belief systems and institutions~

There are far too many Intellergetic Criminals who have mastered the ability to manipulate this cyclic density to their own ends to attract followers, profit, and power~

Once an Energy Signature has truly been immersed into _____ and experienced entangled, fully expanded space, like our Horace on the beach, they are fundamentally changed in terms of perception and harmonic resonance~ Sensations of the changes in one's sphere of being show up as glaring and uncomfortable~ Intellergy becomes aware of the tensions and limitations of returning to Qen4 awareness~ If one could observe the Signature at this time, they would see a lightness in certain density centers as these accumulations begin to dissipate~ One is now on the Qensional Bridge, embarking upon a transitional path to higher frequencies of perception that is all but impossible to resist~

An Energy Signature with introductory Signome realizations will also soon find that their newfound harmonics do not necessarily resonate as well with previous relationships or environments that had previously been comfortable and harmonious~

For example, let's say that one had a favorite restaurant where they went regularly to meet friends, eat, drink, and socialize~ Here they have always felt welcome, accepted, and free to *be themselves*~ However, after returning awakened from a Qen5 state, somehow the frequencies in the environment of the restaurant seem to be slightly discordant~ Conversation with friends and coworkers seems to resonate more as dissonance, irritating certain energy centers that had never before responded in such a way~ After a time, the Energy Signature finds itself going to the restaurant less and less, finding excuses not to join this group of friends as much as before~ This newly awakened Intellergetic being begins to pick up and read materials written by enlightened Bearers and search for others with whom they might share their experience and ask questions~ Nothing will ever be the same~

Qen4 Intellergy is consumed with the collapsing of wave functions destroying the probability clouds of XPotential that is immersion in _____~ Energy Signatures absorbed in this realm commit their manifestations to the perpetual intention of categorizing, measuring, observing, quantifying, and judging their illusory realities~ This attachment is ultimately fear-based~ It is the fear of loss of identity, of objects, of belief systems, of preconceptions, and of their concept of reality itself~

~~~*~~~

For a slight diversion, let us put forth a little mind experiment or Thojourn to describe the next few seconds for the average human:

> *In the NOW present moment, a reactive perception is made that could be an observation, a measurement, or a quantification of an object, event, or phenomenon~ For this Thojourn, let us use the experience of watching a shooting star crossing a clear, summer night sky~*
>
> *The infinite potential of the Probability Cloud has disappeared at this observation of the meteor and the collapse of the Wave Function~ This perception is instantly absorbed, processed, and stored in the Intellergetic Data Bank of the Energy Signature as a density that has resonated harmonically with a particular center of sympathetic frequencies collapsing every Wave Function in its flow~ For many, the processing of this experience would be one of wonder and beauty~*
>
> *At this time, a reaction is stirred, resulting in some kind of intelligent energy transmission in the form of an action, word, deed, or thought~ The reaction could take the form of a memory of another time seeing a meteor streak across the sky, perhaps in the company of a friend~ Perhaps, it would simply be "Wow!"*
>
> *Even if the reaction is perceived to be conscious and mindful in nature, objective reflection will almost always reveal that the response was the result of accumulations of experience, expectation, and derivative, formulaic, and predictable belief systems~ Those of us with the Human Disease take these processes completely for granted as*

*normal life **learning to trust our instincts or having faith in our beliefs**~ These are the illusions~*

In the end, the entire experience was anything but spontaneous and mindful~ It was the completely reactive response of a programmed human robot~

~~~*~~~

Psychologists speak of those with control issues~ The Human Disease is the definitive control issue~ Humans seem to thrive on controlling the complexity of existence~ The more definition and density one is able to compact into one's Energy Signature, the more secure they appear to feel~ This, of course, is only true until the inevitable entropy of all density systems begins to deteriorate these realities, and then fear becomes more and more compounded in a continuous spiraling cycle of destruction and reconstruction~ This rotation of phases is the *squirrel cage* of the Human Disease that should be completely unnecessary from an expanded consciousness perspective, and yet humans come to expect and live from them~

"Roll with the punches," they may say~

"Just another bump in the road," says another~

"Man, life sure threw me a curve that time."

All of this complexity was constructed to divide, quantify, and define what is, in the end, simply one infinite pool of limitless energy and consciousness~

Blindsight strives for the simplest, most expanded perception of consciousness and energy~ All attempts to classify _____ can only result in accumulation of densities leading to turbulence, complexity, duality, and attachment~ In our discussion below of the I~Me~My~Mine phenomenon, we will present the concepts as a reflection of the Energy Signature condition~ For the purposes of the present discussion, we feel it is important to view each of the I manifestations in an Intellergetic being as *crystallized* Energy Signatures that have their own set of informational energy densities~ These *I's* often develop further crystallizations within them, also with corresponding concentrations~ An Energy Signature not focused on multiversal immersion in _____ is virtually trapped in a vortex of conflicting identity forces that become all but impossible to disperse as they continue to feed on their energetic *tails* until the ultimate dissolution of the manifestation~

In the Chinese tradition, *Qi* is a word for life force, loosely translated as *material energy* and used often in healing practices, spirituality, and martial arts~ Qi has many similar references in cultures all over the world, ranging from the *ruah* of the Jews to *mana* in Hawaiian culture, to even *The Force* of the Jedi in Star Wars~ Early Chinese thinkers believed Qi to have different qualities that could manifest as liquid, solid, and as mentioned, the life force~ Qi can be likened to Intellergy, referring to an intelligent flow of energy that circulates around and through an Energy Signature~

> *For my ally is the Force, and powerful ally it is~*
> *Life creates it, makes it grow~ Its energy surrounds us*
> *and binds us~*
> —Yoda

In many ways, the practices in observance of Qi, along with other energy practices, act as sorts of *stepping stones* to more expanded awarenesses~ For some, these works can be like a bridge to Qen5 intelligence, and as such should be pursued if this is one's evolutionary point of resonance~ Should one engage in one of these practices, the main pitfall is always attachment and limitations with its accumulations of density~

Blindsight would offer that all Energy Signatures exist in what they perceive as Intellergetic density states unique to themselves alone~ These states consist of resonances and harmonics attracting the information and energies necessary either to reinforce their Human Disease or to aid in the transition to higher consciousness~ In the end, it is most important to remember there are ultimately no individual Energy Signatures~ All the illusory objects, actions, thoughts, and beings are entangled in the One Mind that is the observational aperture of

_____~

The cure for the Human Disease lies in the intentional inoculation of one's self against any and all temptations to qualify, judge, or attach to language or imagery relative to consciousness or energy~

> *Healing begins in the moment that we are still and*
> *recognize that something must change~*
> — Kevin Hall

Our woman, described in the chapter on the Human Disease, was living a relatively comfortable and confident life, completely unaware of the disease that was slowly ending her~ As Energy Signatures infected with the Human Disease, how many of us can say the same?

Many Intellergies who possess density propensity for a religious construct hold belief systems allowing for the existence of a Devil-being as the opposite of God~ One should look no further for Luciferian influence than the Human Disease~ Intellergy becomes like the hamster on the wheel, ever in pursuit of the next perception, with no ultimate destination other than the sensation of motion and the passing of time~ If there were a Satan and he had a brilliant tool for temptation and deception, that implement would be the Human Disease~

> *The greatest trick the Devil ever pulled was convincing*
> *the world he doesn't exist~*
> — Charles Baudelaire

One may view Gurdjieff's language of *slavery* as too graphic; however, for us this is an intimate reality~ We constantly struggle against the chains of Qen4 illusion~ Once one has experienced one's true energetic nature and the possibilities of immersion in _____, it is all but unbearable to settle for existence as a prisoner in a manifestation built on illusion~

> *It took me four years to paint like Raphael, but a lifetime*
> *to paint like a child~*
> — Pablo Picasso

Perhaps the most difficult aspect of Qen4 awareness to escape is the incredible depth of beauty and heart permeating our manifestation at these denser frequencies~

Think of a West Texas sunset, Bridal Veil Falls after a week of spring rains, puppy breath, a lightning flare, the flash of a blade, the first four notes of Beethoven's Fifth, the sweet release of a lover's sigh, and every other nuance of color on the palette of Qen4 creation~ This vast collection of splendor, passion, and inspiration can prove to be some of the most difficult densities to dissipate~ Many Intellergetic Beings have manifested remarkably successful and beautiful existences with defining attachments to the associations, objects, and perceived accolades that come with such a life~

Once there has been a full immersion into Qen5 awareness, most Intellergies will begin to see the attachment to these aspects begin to dim; however, it will be the lingering resonances of these pleasure and recognition harmonics that will continue to lure the being back into Qen4 tension/release cycles~

Even now, and most likely for a very long time, we personally will be facing the challenges of a Qen4 Energy Signature with Qen5 resonance~ Please do not mistake our confidence in writing this book on Blindsight for mastery~ We are far from masters of our Intellergy~ Our goal is to share our path, which engages far more mystery than mastery in our struggles, for others to hopefully find some inspiration in these perspectives~

I feel like a bird in flight trying to understand the air~
— MWUSO

Addiction to Tension/Release Intellergy~

This is the third aspect of the Human Disease~

The One Mind and the Multiverse are constructed of illusory dualities such as AC/DC current, light-dark, creation-destruction, particle-wave~ It is no wonder that human Intellergy would develop such propensities and susceptibilities; it is intrinsic in our Intellergetic nature~ From the earliest artwork of the Neanderthal culture, tension/release energy has been infused in depictions of good versus bad spirits or the pressures of the hunt versus the relief of the kill~

Take, for example, the commercials in the media of the millennium~ It would appear much simpler to sell aspirin if one has already sold the headache~ Observe the person with simulated agony from the pounding in their skulls, only to reappear in the bliss of relief 20 seconds later after two little tablets and a swallow of water~ If one wants to sell medication for erectile dysfunction . . . well, we think you get the point~

The most disturbing and unfortunate result of the accumulation of density with regards to millennium art, literature, and media is that it has begun to condition Energy Signatures to crave this type of tension/release stimulation in their personal Intellergies~

Very subtly, humans have been conditioned to need some sort of constant progression of activity, energy, or information, and the Terra Signature has witnessed a virtual explosion of maladies given the names Attention Deficit or Hyperactivity~ The acceleration of input has also generated other responses, such as Sensory Deprivation Syndromes, Depression, and Pervasive Developmental Disorders~

This is simply an addiction~ One of our favorite definitions of addiction is *a condition characterized by compulsive engagement in rewarding stimuli despite adverse consequences*~ From a Human Disease perspective, it appears that it is becoming progressively more difficult for Energy Signatures to maintain relaxed, present, focused intention for long periods of time~ Phases of inactivity or lack of stimulation engenders inflammation in density centers that actually engage addictive Intellergy Forming to create manifestations filling these uncomfortable voids~

Reactive Nature~

This is the fourth symptom of the Human Disease.

For every action there is always
an equal and opposite reaction~
— Isaac Newton

Knowing how the environment is
playing you and pulling your strings
is critical to making responsive
rather than reactive moves~
— Ronald A. Heifetz

Life is 10 percent what happens to you
And 90 percent how you react to it~
— Charles R. Swindoll

The *Fight-or-Flight Response* was first identified by Walter Bradford Cannon~ He observed that the sympathetic nervous system in animals reacts to threats of survival with a flood of hormones preparing the animal to fight or flee~

For some humans in these scenarios, there is the ability of Emotional Regulation~ This is a capacity for being proactive in the presence of a threat or traumatic event that could result in a calmer and more appropriate response~ In others of us, there is Emotional Reactivity that could lead to anxiety or aggression~

All of us have felt these types of reactions many times over the course of our manifestation, so it would seem the best way to call attention to this symptom~ However, in the Human Disease, our Reactive Nature is much more subtle~ It is not always so apparent and flooded with hormonal responses~ Reactions are rarely a function of fight-or-flight, and our lives are a constant progression of reaction after reaction~ These can range from bodily functions such as reactions to temperature or hunger to those of reacting to traffic while driving or carrying on a conversation~

Blindsight is not so unrealistic as to believe we must cease all reactions as a path to a cure for the Human Disease~ What it encouraged is simply a more mindful and present awareness of when and how we react to the energy and intelligence around and inside of us~ With just the smallest amount of conscious observing of our little *overreactions* to life's little bumps, it will very quickly awaken the observer to the many small, unconscious responses that mold our realities and our futures~ Our Observer can also give us the wonderful opportunities we have to intervene with mindful attention and intention to Intellergetically form the next pieces of our potential~

Never let reality fool you into believing that it's real~
— MWUSO

Humanity has come to admire those of us with *great instincts*~ We have come to *trust our gut* on certain major decisions~ These traits are ingrained in us as a result of thousands of years of survival~ There is a perception of control that comes with these qualities~ Often one's beliefs about their abilities proves to be either an underestimation or an overestimation of their actual capacity~

The questions for us become:

At what point do our gut and our instincts get in the way of what might be better, more efficient solutions, decisions, or actions?

What would be our potential if, in every waking moment, our responses to life were conscious, present, and proactive rather than reflexive and habitual?

Most of us have encountered that person who is having a bad day, and we just happen to be the one that is in their path~ This Energy Signature has most likely had many of their recessive densities inflamed by an event, an encounter, or possibly just their own mistake relative to past experience ~

It is at this point for most of us that our fight-or-flight response kicks in~ However, any emotional reactivity of this kind will usually result in an escalation of this person's anxiety and aggression, as well as our own~ The outcome is most often some type of confrontation or argument~

The Human Disease can't seem to allow someone to just step into our lives with all this hostility and anger, expecting us to just stand there and take it~ Right?

We aren't really sure when it came upon us *(and it may have something to do with those early out-of-body and near-death experiences)*, but whenever we are met with someone who is hostile against us for any reason, we become what we have come to call the *observer in the conversation~* With this distance and perspective from the event unfolding in front of us, we always seem to *see* the pain and the confusion in the being, not the angry, threatening person~ This has an amazing calming effect on us as we allow their aggression to blow through us like wind through the grass~

Over a lifetime as a performing musician with decades in bar bands, we have been shot at, had knives pulled on us, been threatened with attacks, and more than once placed in danger of physical harm because we didn't play *Sweet Home Alabama* soon enough for the third time of the night~ On each of these occasions, there was simply no fear, only acceptance and composed presence~ It is our experience that even the most scared or aggressive individual is not able to sustain their aggression for long if there is no fear response to fuel it~ It seemed to work out great for Daniel in the Lion's Den~

~~~*~~~

For many years, we worked as a solo performer with just a piano, a drum machine and a microphone~ Our style was mostly on the side of southern rock, boogie, and blues~ We enjoyed playing loud, making the house jump, the dance floor bounce, and the cash register ring~

At one time, we were invited to play at an out-of-the-way bar, back in the windfarms up in the Altamont Hills of California~ In the afternoon, we drove into a parking lot filled with pickup trucks and a bar lined up with fishermen, blue-collar workers, and a couple random physicists from the nearby Livermore Labs~ We set up our gear, then left to find some dinner and clean up before returning for the gig~

As we drove into the lot that night, all of the pickup trucks were gone and had been replaced with dozens of Harley-Davidson motorcycles~ As we entered a room of black leather, chains, and tattoos in our jeans and boots, we couldn't help but recall the words of the great and humble George Gobel, finding himself on Johnny Carson's stage, seated with legends Bob Hope and Dean Martin, "Have you ever felt the whole world was a tuxedo and you were a pair of brown shoes?"

We made our way to the stage fully confident that when we broke into our Bob Seger, Creedence, Doors, Little Richard, and the rest, all would be well~

We had not even played one note when, from the back of the room, we heard, "Hey, play the 'Midnight Hour'!"

We remember smiling and saying, "No problem, bro~ That one's pretty strong~ Let us warm up the ol' voice a little, and we'll be happy to do it for you."

"Play the goddamn 'Midnight Hour' right now or I'm gonna kick your ass," we heard rumble through a room that had gotten surprisingly quiet~

Remember, to this point, we had not played a single note~

"Ladies and gentlemen, we're going to take a short break," we heard our self say into the mike~

We left the piano and walked straight to the back of the room to find a giant of a biker in full leathers and a face that was mostly beard with a large Bowie knife scabbarded down his leg~

"Hi, I'm Mikal," looking up into his fearsome face~ "Just wanted you to know that I play requests all night long, but I do not play demands." And then I stuck out my hand and gave him my widest Texan grin~

We watched as his resolve began to crumble and within seconds, he took my hand and said," You know, little fella, you're all right~ Lemme buy you a beer~ By the way, the name's Booger~ Next time your finger's up your nose, think of me."

We played this biker bar every weekend for almost three years~

~~~*~~~

It has been said that all human reaction can be separated into either love or fear~ Fear is density and, as such, it cannot endure~ In our moment with Booger, our love intention in the moment completely dissipated his fear-based densities~ As you are met with all the many obstacles, confrontations, and even threats in your manifestation, strive to be the calm *Observer* in your life and the mindful, loving responder~

Here in the millennium, most of us rarely engage our Fight-or-Flight response in the ways our distant ancestors did with regards to environmental and predatory threats~ However, there is something that has manifested in this age that has placed us all on a different kind of alert, and this is the advent of the smartphone and interactive digital devices~

TwittaGoogaFaceChat

TwittaGoogaFaceChat is a term we play with to convey the labyrinth that is social media in this early millennium~ Through the *TwittaGoogaFaceChat*, someone is trying to contact you at all times, twenty-four hours a day, seven days a week~ Our inherent nature is that of a social being, and in that spirit there is little more compelling than our connections to that network~ Proven to activate our mind/brain's reward systems, we are not only wired to respond to input of a human nature, but also just about any new or interesting sight or sound~

TwittaGoogaFaceChat sends out that sudden beep, chime, or ringtone du jour with its promises of any sort of stimulation to drag us away from whatever intention or activity in which we may be engaged~

A. T. Jersild, in 1927, was testing humans' abilities to use what he called *Task Switching* as a kind of *cognitive flexibility*, allowing one to adapt rapidly and efficiently to new environments, problems, or threats~ Integral to survival as well as success, there is still something that is called a *Switch Cost* when a task or thought is disrupted by reaction to a digital alert~ The price paid is that of slowed performance, decreased accuracy, and reinforced addictive attachment~

Many of us are better at multitasking than others, but it is an irrefutable fact that the truly great inspirations, inventions, insights, and epiphanies find their roots in single-minded focus, attention, and mindful intention~

In a study found in the peer-reviewed science journal *PLOS One,* it was discovered that those who spend a large amount of time *media multitasking* and juggling apps, websites, and other digital stimuli actually tend to have less gray matter in their brains associated with the control of thought and emotion~ Ultimately, this process is progressively programming humans to be more reactive and robotic~

TwittaGoogaFaceChat is most successful to its own ends when we blindly react to this barrage of Intellergy~ They find more hits on their sites, more subscriptions to their apps, and more purchases of their products and services if we simply and robotically react~

At the same time, *TwittaGoogaFaceChat* is the singular most compelling vehicle for the bringing together the higher frequency Intellergetic beings literally exploding in numbers in the Terra Signature at this time as we approach the Quantum Point of the Great Shift that is occurring at this time~ It is the conscious Intellergetic use of this globally networked brain that is fueling and energizing the portals and pods of Light Bearers~

We can never forget that an Energy Signature is a reactive being~ We have been reacting to the expansion of this Multiverse since the Big Bang and the Inflationary Epoch~ The Qentinuum carries us in its torrential flow, but as Intellergy infused with One Mind and _____, we possess free will and choice~ These are our inoculations against the Human Disease~

~~~*~~~

G. I. Gurdjieff related a parable of sorts about a horse-drawn carriage~

In his scenario, the carriage is drawn by a team of beautiful, spirited horses with a driver~ The Master is riding inside~

The Horses represent emotion, passion, and energy~

The Driver symbolizes the mind~

The Carriage implies the body~

The Master is our true nature~

If the Driver does not maintain the body of the carriage in good repair, it can fail, injuring the Master, the Driver, and the Horses~

If the Driver cannot control the emotions and spirits of the horses, they could run away with the carriage, again endangering all~

If the Driver is too self-consumed or undisciplined, he will drive the carriage recklessly and without purpose, never reaching a true destination~

When the Driver is mindfully skilled and present, the Carriage is sound and beautifully maintained, the Horses are strong, healthy and responsive, and the destination is secure~

The Driver, unencumbered by willful emotion and distractive reaction, will now be able to hear and carry out the true intentions of the Master ~

~~~*~~~

Please begin *Now* to direct attention and intention to the awareness of your participation in **Language, Attachment, Addiction to Tension/Release Intellergy,** and **Reactive Nature,** and accept the challenges that appear as paths to freedom from the Human Disease~

The natural healing force within each one of us is the greatest force in getting well~
— Hippocrates

Physician, heal thyself~
— Luke 4:23

We are not physical machines that have learned to think...
(but) thoughts that have learned to create a physical
machine~
— Deepak Chopra

If you, the reader, are not seeing the impact and effects the Human Disease is having in your life; if you are not shocked and angry at how far you have come, being led around throughout your life by reactive, predictable, and formulaic intention; if you have not awakened to a new and abiding hunger to change and evolve from this realization, it may prove best to cease reading at this point~

Our personal awakening to the reality that the nature of the densities attracted to our Energy Signature had been constructed of completely false Intellergy became impossible to accept~ It is our sincere hope this reality is untenable for you as well~

What are we supposed to do now?
How will we ever be able to escape?
Is there anyone out there to help us?

It may sound as though we are calling for a revolution~

For us, a revolution is exactly what it has become~ The intimate realization of our mindless enslavement to reactive existence that Gurdjieff spoke of incited a passionate rebellion against all of our belief systems~

Please choose now, with mindful, passionate intention, to escape the confines and illusions that have enslaved you~ We are not just robotic bags of skin and bone reacting to the programming of our past beliefs~ We do not have to continue answering to the preconceptions and expectations that have accumulated as densities in our Energy Signature~

We are self-aware, intelligent energy~
This is our truth~
It is our free choice to live from this truth~
Anyone existing as a physical brain body
Has the same truth; the same choice~
— MWUSO

NOTES

Bholcano by Kelci Auliya

IV

I~Me~My~Mine

I am not a physicist~
I am a father~
I am not a cosmologist~
I am a musician~
I am not a neuropsychologist~
I am a husband~
I am not a philosopher~
I am a brother~
I am not a scientist~
I am an author~

Isn't it simple to use *I am* in reference to everything we say and do?

There are so many *I am* statements and points of view in our existence~ I am black~ I am Republican~ I am Muslim~ I am a woman~ I am Chinese~ I am~I am~I am and on and on~

If one is truly honest with oneself, they will soon see there is really no solitary action in their existence that is solely their own~ Even the simplest acts of the senses are reactive, stemming from accumulations of past stimuli, impressions, expectations, and beliefs~

One might say, "I see a dog." This is an act most humans take for granted~ The Energy Signature is the instigator, the receiver of the information, the processor and the transmitter of the experience~ In reality, many other people, animals, and prior events went into the ability to recognize these things and express them~ There must have been a parent, a teacher, or a friend who first showed them the dog, identified it, named it, and taught the word~

Who first saw a dog and called it a dog?

What actions by others followed that made the word, *dog*, commonplace and accepted?

Many is the time, for example, we encounter employees such as salespeople, warehouse workers, weather reporters, or whomever in the business world referring to their work environment or their actions as solely their own~

"All of *MY* computers are on sale today."

"*MY* trucks will be able to deliver this afternoon."

"*MY* weather forecast is coming up at 11."

These professionals might even offer that they are taking ownership of their positions, and in some cases, they could be commended and rewarded for this perspective~

Blindsight would suggest the question of how different these thoughts would convey if presented from a "We~Us~Our~" position?

"All of _OUR_ TVs are on sale today."

"_OUR_ trucks will be able to deliver this afternoon."

"_OUR_ weather forecast is coming up at 11."

This expanded language instantly conveys the idea of a team supporting the customer or the audience and reflects upon the speaker as a larger, more aware, entangled being~

For this simulation, this is why we write from the *We* perspective, with respect to the concept of what physicists call Quantum Entanglement or the principle that all perceived objects, ideas, or entities in this so-called Multiverse are inextricably infused within each other~ The use of *we* implies an expanded beingness that is not defined by a name, body, or position in space-time~

At this juncture, we would like to challenge the reader to begin observing themselves with calm indifference every time the word *I* is expressed in their conversations and thoughts~

We issue this as a challenge because, if taken seriously with objective examination, most will find this exercise daunting, as well as overwhelming at first~ The I~Me~My~Mine~ imbedded in every Energy Signature is extremely crafty and elusive~ Oftentimes, one will find themselves observing one of their *I*s only to discover they are using a different *I* in the observation that is filtering, altering, and quantifying the information gleaned from the other *I* to its own ends~

For most, it is very difficult to remain objective with their innermost identity structures~ If one is honest in their observations, one will begin to see that each time *I* is employed, it triggers a *charge* in the Energy Signature, exciting a density or collection of energies associated with an experience, a sickness, a trauma, a vivid impression, a relationship, a victory, a defeat, a memory, an inspiration or others~

One will soon begin to observe in themselves that there is a happy *I*, a sad *I*, an angry *I*, an anxious *I*, a tired *I*, a focused *I*, and on and on~

If one can learn to become honest and impartial in these observations, one will have to admit that the presence, energy, intelligence, and expression of a tired or angry *I* is far different from that of an inspired or enthusiastic *I*~ Isn't it curious that those with the Human Disease have become comfortable with all these entities living inside their Energy Signature, rising to power at any given moment to gain control of Intellergy Forming realities to which other entities must rise in reaction to forge other actualities for others? These are cycles of *crystallization* in the energy field~ The Energy Signature becomes comfortable with a *personality* that consists of the collective Intellergy of all these *Is*~ This *Identity Assemblage* becomes perceived as just one global Energy Signature under one name, body, position, and mind/brain~

This is a little concerning, don't you think?

It is also important to note that, from a Blindsight position, each of these *Is* is also its own Energy Signature, with unique and individual traits, goals, actions, and reactions~ An angry *I* might react in an instant from a perceived threat or attack without any proactive forethought~ This anger could manifest as a returned verbal assault or perhaps even striking out physically~ A calm, centered *I* would most likely counter in an entirely different manner~

How many times have we heard or said ourselves, "I was just mad; I didn't mean what I said?"

Believe us when we say that the angry *I* meant every word~ Its Energy Signature was reacting to inflamed density accumulations and transmitting accordingly~ In this context, are we true to ourselves and others if we are able to so easily dismiss the thoughts and actions of internal Intellergies within us? Taking responsibility for all aspects of one's manifestation by engaging our Observer with playful intention is the fundamental first step toward awakening~ Intimate, brutal, beautiful, fearless, honest, ruthless, and wonderful observation of one's universal Energy Signature with all its centers, densities, and imperfections grows over time into a magnificent process~

All of these Intellergies coexist within the global Intellergy, proceeding to harmonize and resonate like the strings on a single guitar~ *(Later, we will offer a Thojourn called the Multiversal Guitar to expand on this idea~)* For the guitarist, each string can be played singularly or in concert with any combination of the other strings, creating harmonies and chordal structures~ An observer might comment they are seeing and hearing one guitar without recognizing or listening to the contributions of the individual strings or the resonances produced within the body of the instrument~

The same can be said of an Energy Signature made up of many different I~Me~My~Mine Intellergies~ The common observer may view this Energy Signature as a single body/mind personality, as in the guitar metaphor, while in truth, what they observe is a swirling probability cloud in constant flux, cycling at near light speed and obliterating wave functions in, out, and between this Multiverse and other informational energy systems~

There is but one true conscious and present I~Me~My~Mine act and that is immersion in _____ ~

A man once said to Buddha, "I want happiness."
Buddha said, "First remove 'I,' that's ego~
Then remove 'want,' that is desire~
See, now you are only left with happiness."

~~~*~~~

Mir by Erin Nicole

V

Breathe~Play~Now~

The only cure possible for the Human Disease and our attachment to I~Me~My~Mine Intellergy begins within our own mind/brain, where we are *(Here)* and when we are *(Now)*~

Throughout all recorded human history, there have been many practices designed for exploring, calming, and centering the mind~

The oldest records of these disciplines is that of the Hindu, going back as far as 3,300 BCE, with its mantras, yoga, and meditation, although they most assuredly were in practice long before~ Under a fig tree, Gautama resolved to sit in contemplative meditation until he had discovered a solution to human suffering~ In Judaism, there are certain sects that engage in an informal practice of mindfulness before prayer~ Many use words such as *eschad,* meaning *one,* just as a Hindu might use the word *Om* as a mantra and others use a visualization call a *shiviti,* much like a mandala for contemplation and meditation~ Contemplation is one of the early cornerstones of Christian mysticism and is invoked to achieve unity with God~ It is to bring one's entire life into a place of listening and learning~ Meditation is central also to the Muslim faith~ Similar to yoga, with its physical sequence of standing, raising arms, bending, lying prostrate, and kneeling~ The whirling dervishes of the Sufis believe this ritual brings them closer to God~

True centering down transcends worship~
— Phillip Gulley, *Living the Quaker Way*

Blindsight compels that some form of regular practice of intentive mindfulness must be engaged for the release of the densities of the Human Disease and the raising of perception frequency harmonics~

Only loving kindness and right mindfulness can free us~
— Maha Ghosananda

Mind*ful*ness (noun)

The quality or state of being conscious or aware of something~

A mental state of focusing one's awareness on the present moment while acknowledging one's feelings, thoughts, and bodily sensations~

Used as a therapeutic technique~
—Oxford Living Dictionary

When you love someone, the best thing you can offer is your presence~
How can you love if you are not there?
— Thich Nhat Hanh

Breathe~Play~Now is not a Thojourn with any kind of destination, although it is a most excellent vehicle for facilitating the very best mindful experiences~ It is a practice so amazingly simple that a child can master it with ease~ Blindsight offers it as a place to begin, and it should be used before and during any creative visualization, intention, or meditation~

We have intentionally placed this chapter here in hopes you will take a short pause after reading to experiment with this wonderful mindfulness to carry you through the rest of our simulation~

Breathe~Play~Now is a little different from most other meditation and contemplation techniques, with its addition of the element of *Play~* The dynamic of *Play* adds a completely new dimension to the experience that is fresh and spontaneous~

Breathe~Play~Now can be used anywhere, anytime~ It can be done sitting quietly, in yogic practice, walking, waiting in line, driving, working, or any activity~ It is an amazing exercise right before sleep or upon waking to relax the mind, boost the autoimmune system, and focus the consciousness~ Used only once during a chaotic day, it is a great way to reset your Intellergy in more peaceful and mindful ways~ It is even safe to use while operating heavy machinery!

Using this exercise for about ten minutes in the morning when you first wake up energizes the body and brain with increased oxygen levels while centering and calming the mind to begin Intellergy Forming the next moments of your day's reality~

When employed at bedtime, the increased oxygen will naturally release melatonin in the brain to speed and deepen relaxation, while at the same time once again centering the mind for an Intellergetic transition to rest~

Use it often during the day~ Even if it is only one repetition, you will immediately notice a shift in Intellergy and presence~ If you find yourself in any kind of stressful situation or a place where an important decision or reaction is required, *Breathe~Play~Now* is an amazing way to instantly focus the mind and relax the body~

Breathe~Play~Now consists of three aspects, but they all work in concert~ No one piece is more important than another, and when employed with mindful, self-aware intention, the results are magnificent~

It all begins with the **Breath**~

Breath is the Intellergetic nexus of your being~

Breath is the point of interconnection for the Energy Signature, One Mind, and _____~

When breathing involuntarily, Intellergy is in a reactive mode, unconscious relative to multiversal intention~

By consciously controlling the breath, one engages the full potential of present, aware evolutionary exercise~

How many humans in waking consciousness have taken a full, diaphragmatic breath with mindful intention in weeks?

Months?

Ever?

How long has it been since you took one long, full, wonderful breath on purpose?

How long since you took ten of them in a row?

Have you ever done it for thirty minutes straight?

Go ahead, take one now~

Instead of just filling your lungs, use imagination and Thojourn intention, letting your inhalation expand, filling with the true nature of your infinite Qen Intellergy~

As you exhale, envision all the energy, intelligence, power, creativity, and love offered by One Mind being drawn back into your Energy Signature~

May we offer you another?

The true man breathes with his heels~
— Chuang Tzu

He lives most life who breathes most air~
— Elizabeth Barrett Browning

Breathe~ Let go~
And remind yourself that this very moment is the only
one you know you have for sure~
— Oprah Winfrey

Breathing is the only bodily function a human can immediately control~ One can speed up the breath or slow it down~ One can hold it or release it~ How one is breathing can be an indicator of a mood, such as anxiety, concentration, or sexual arousal~ Breath can reflect physical activity, such as exertion or preparation for an endeavor~

Deep breathing has been medically proven to increase brain power and cortical thickness~ It improves heart rate variability, lowers stress levels, and promotes positive outlooks on life~ Regular deep breaths lower blood pressure~ They decrease anxiety and increase concentration~

Go ahead, take another one~

Great yogic masters engaged in *Pranayama,* the formal practice of controlling the breath and focusing the energies of the universe within the body and mind~ Pranayama has been employed for centuries, and many of these masters have used the training as a platform to reach the deeper control of other bodily functions such as heart rate, blood flow, body temperature, and pain management~

*** *Please understand there are right ways and wrong ways to use Pranayama~ If it is your choice to pursue this path, there can be negative effects~ Make yourself available of an experienced teacher if possible before experimenting with the techniques~* ***

Ultimately, the regular practice of deep breathing can be truly life-altering, and it is the first piece of the basic *Breathe~Play~Now* Blindsight Meditation~

Please take another at this time~

~~~*~~~

As you proceed with this exercise, breathe not just to fill your lungs and expand your diaphragm, but use Thojourn abilities to send the intention of your breath into infinite, fully expanded space~ Use the present awareness of your breath to embrace your true eternal Qen nature~ Engage the Qen5 Meta-Intellergy to enhance awareness of the process~ As you exhale, download the unlimited energy and potential from One Mind, allowing it to fill your Energy Signature, sweeping away density accumulations~

In concert with Play and Now Intellergy, this complete expression of breath, when used consistently, becomes a simple, beautiful way to connect with Multiversal Intellergy and _____~

The second aspect of the exercise is **Play**~

Play is the only way the highest intelligence of humankind can unfold~
— Joseph Chilton Pearce

Don't play what's there,
play what's not there~
— Miles Davis

The true object of all human life is play~
Earth is a task garden:
Heaven is a playground~
— G. K. Chesterton

We speak often of the concepts of *Play* and the infusion of *Playfulness* into present, self-aware consciousness~

Playfulness is a state of mind; it is also a choice~

There are no rules in mindplay; there is no predefined structure~ Play is what it is for you and only you~

~~~*~~~

Leaving on a recent flight to points South, we had parked our vehicle at an offsite lot and jumped on a shuttle for the airport~

Our driver was very professional, quick, and helpful with our bags~

"How are you doing?" he asked, but there was no smile, and we did not feel there was any real intention on his part that he really cared how we felt at the time~

We tipped him $2~

On our return, we got on another shuttle for a ride back to our truck, and this time the driver had a big smile on his face~ Upon reflection later, we recalled that every time we looked at him, he was smiling~

When he asked how we were doing, we replied, "Fine, how are you?"

He said enthusiastically, "It's another glorious day in Parking Shuttle Paradise!"

With his response after hours of travel, airplanes, lines, and airports, we suddenly felt a little lighter in spirit and welcomed back~

We tipped him $5~

This driver had engaged positive intention to infuse play and well-being into his work, as well as our lives~

~~~*~~~

Within the exercise of *Breathe~Play~Now*, we have found the simplest way to engage in playfulness is to put a very slight smile in our closed eyes~ We also induce the feeling of a small *smile in the mind~* When one does this without any inhibition or preconception, one will almost without fail begin to sense a lighter, more expanded sense of joy~

Several years ago, brain researchers used scans to determine the areas in the human brain where happiness resides, and they experimented with many subjects ranging in age, sex, and cultural backgrounds~ They came upon an elderly Buddhist monk who had spent virtually his whole life as a recluse, meditating fully six to eight hours a day~ The centers for pleasure and joy in this old man's brain fairly exploded with color and energy on the scans~ The researchers finally had to admit that they had most probably found *the happiest man on earth~*

Medical research shows that smiling releases the same endorphins as running or jogging, while lowering anxiety and contributing to happiness~ It has been proven that smiling increases the production of white blood cells to fight infections and accelerate healing~ A smiling person has a lower resting heartbeat than the same person with a neutral expression~

Socially, smiles are contagious~ They convey energy that one is more personable, relaxed, empathetic, and trustworthy~ You are more attractive when you smile, and management studies have proven that leaders who smile engender more loyalty and extra effort from their employees and constituents than those who do not~

One last thought on smiles, and that is about your countenance~ One's countenance, in simplest terms, is the look on one's face~ We have ourselves become more sensitive to this as a result of many years of playing our *Searching for Signs of Life* game~ We have spent so much time looking into people's faces that we have become much more aware of the face we present to the world~ If you decide to play the game, start simple at the grocery store and you will literally be amazed at the looks of confusion, seriousness, and anxiety on the faces you pass~ How different would the Intellergy be at the store if everyone was present, in the moment, breathing deeply, and holding a smile in their eyes?

What will be the countenance you present to the Intellergies you encounter?

What is your countenance at this moment?

Smile intentionally right now for a few seconds and just feel the transformation in perception, presence, and consciousness~

What will be your countenance at the time of your transition?

Play with abandon in every moment and observe the remarkable changes in the Intellergy within your Energy Signature~

When engaging *Breathe~Smile~Now~* begin to practice infusing a state of playfulness into your awareness as you breathe deeply~ This is most easily accomplished at first by bringing memory densities of playful past experiences into *NOW* consciousness~ The more skilled you become, the easier it will be to simply engender a feeling of play for absolutely no reason into any moment you choose~

The third characteristic of our exercise is **Now**~

> *Nothing has happened in the past~*
> *It happened in the Now~*
> *Nothing will ever happen in the future~*
> *It will happen in the Now~*
> — Eckhart Tolle

> *If not now, when?*
> — Mikhail Gorbachev

> *Just feel a whole lot more like you do now than you did*
> *just a while ago~*
> — The Tortugans

It seems almost an insult to you, the reader, after you have endured all the densities in these past virtual pages with the multiple references to *presence, mindfulness, the moment,* _____, and the rest for us to elaborate much more on the importance of endeavoring to remain in the *NOW*~ You must know the importance of this position of manifestation~

Any memory or reliving of past happiness or turmoil is only experienced in the present~ It has no temporal reality; it is an illusion~

Any projections for future events, actions, or relationships are only experienced *NOW;* they have no reality and are ultimately illusory~

Hold every moment sacred~
Give each clarity and meaning,
each the weight of thine awareness,
each its true and due fulfillment~
— Thomas Mann

The practice of *Breathe~Play~Now* is meant to blend all these three aspects in a synergetic knowingness of open, peaceful, playful awareness~

Applied in the waking activities of daily actions and interactions, this practice stimulates heightened awareness, understanding, and loving nature~ Intellergetically, it will improve memory, reaction time, focus, attention span, the ability to learn, patience, and decision-making~

When *Breathe~Play~Now* is used in quiet meditation, before going to sleep or upon waking in the morning, it will relax mental activity, increase blood flow and nerve activity~

Most importantly, *Breathe~Play~Now* should be used to quiet the mind, dispelling all thought of object, motion, personality, or action~ Ultimately, the *NOW* aspect of the practice is incredibly important~ It is the presence in the quiet between breaths that _____ resides~ If this space is unclouded by the turbulence of thought, the joy of beingness in fully expanded space becomes possible~ With practice, this presence can be extended indefinitely~

There is very simply no downside to the practice of *Breathe~Play~Now~*

Qen Breath

Please refer to Qen in the Expanded Glossary before attempting this exercise~

And when I breathed,
my breath was lightning~
— Black Elk

Qen Breathing is similar to *Breathe~Play~Now~* with respect to its techniques of deep, mindful, and playful breaths; however, it becomes a more advanced practice by engaging Blindsight Thojourn imagery~ In this exercise, one must begin with the complete, intimate knowing and acceptance of their true nature as infinitely entangled intelligent, energetic beingness in endless Qen space~

Creatively imagining one's Energy Signature at one with Qen, the unending field of Xpotential energy and intelligence, every inhalation originates as the singularity that is the *Impersonal I~* With the expansion of the lungs, one's Intellergy is known as immersed in oneness with this Qentinuum~ All Multiverses emerge from and are contained within this breath~

The following exhalation draws into the Energy Signature all sempiternal Intellergy, filling the beingness with unconditional love and joy~ There is an irrepressible *smile* in this conscious suchness~

With intentive, playful practice, Qen Breathing will raise the Intellergetic frequency responses to the highest resonances and harmonics~ More and more often these timbres will fade, leaving only faint radiance and the pure unmoved Xpotential of intelligence and energy~

There is one last thought we would offer~ There have been literally centuries of abuse, confusion, and even mockery with regard to terms such as those employed here~ Words such as *unconditional love, peace, joy, play,* and the like have become commonplace to the point of desensitization for far too many Energy Signatures~

Please do not let the misunderstandings of the value of knowing your truest Intellergetic nature deter you from finding ways to take these Thojourns and engage in these meditations~

Trust that the realizations and experiences are fundamentally life-altering, not only personally but also from a global standpoint~ The power that is available to you is infinite and unlimited~

Be loving, fearless, present, and playful~

Oh yes, and breathe!

Smile, breathe, and . . .

> *believe in magic~*
> — Mickey Mouse

VI

Outside In

This is a simulation about metamorphosis and evolution~ It is not a reactive process as such, like the transformation of a chrysalis into a butterfly~ Blindsight explores the possibility of conscious, mindful, proactive choices to evolve; it offers opportunities to exist as Intellergetic Beings that are larger, more expanded, richer, and more aware of our true infinite nature~

How do we know we exist?
How do we know that we KNOW we exist?
What existed before the Big Bang?
What exists outside of this Multiverse?

Timothy Leary's dead~
No, no, no, no,
he's outside looking in~
— Moody Blues, "Legend of a Mind"

It is apparent to us that virtually all millennial scientific disciplines involve choices, observations, and intentions that examine the *known universe* from within the universe looking outward~ These brilliant scientists have become consumed with the examination of galaxies, black holes, tetraquarks, sparticles, gravity, Dark Energy, and the like~ Others pursue the mysteries of DNA, microbiomes, or the neural, chemical pathways of the brain~ Some look to the past in bones and caves, writings, oceans and geodesic layers~ Many enthusiastically celebrate the elegance when this universe responds so readily to mathematical equations and theorems~

It continues to baffle the Blindsight mind when these scientists simply ignore the Intention-infused energy and intelligence that preceded all these manifestations~ They are comfortable settling for these Qen4 solutions, discounting all possibility and potential from an Intellergetic perspective~

Blindsight prefers observations that occur from outside looking inward~

All this is done within the context of a quantum mechanical imperative compelling that there are multiple and most probably infinite other Multiverses and dimensions~ For example, another *Multiverse* may manifest gravity as light whose speed is many times that of ours~ In an alternate dimension, self-aware Intellergy might exist as a silicon base rather than carbon~ It is more likely these dimensions bear little resemblance to any conceptual possibility stemming from our experience~ From a multiversal perspective, most of the realities the scientific community considers based in fact are ultimately very limited in perspective~

The explorations of the majority of modern science appears to Blindsight as likening a scientist dedicating their entire career to the understanding of the fire ant with its complex communities, architecture, societies, and biology, while completely excluding the possibility that there might be black ants, red ants or others~ For us, it would appear the data is incomplete~

As of this writing, the tenets of Quantum Mechanics are being challenged and often overturned almost daily~ With all the excitement recently over the confirmation of the Higgs-Boson and the Higgs Field, now there is compelling evidence of phenomena certain physicists are calling *Particle X* and *tetraquarks* that hint at still another connection with the more subtle aspects of the physical Multiverse~ Dark Matter and Dark Energy remain elusive~ A new study is suggesting a fifth force of nature in addition to gravitational, electromagnetic, the strong nuclear force and the weak nuclear force~ These scientists theorize this new force may be emergent as a result of something they call a *protophobic X boson*~ On the other side of the fence, some alternative new math is looking at gravity no longer as a fundamental force, but rather as an *emergent force* that may be created from interactions with other forces~ Tune in tomorrow when the quantum tables are turned over once again~ Some Big Bang mathematics are now allowing for a *Big Bounce* occurring when a prior universe collapsed upon itself through a singularity emerging as this completely newly formed Multiverse; however, in the face of Dark Energy, this is highly unlikely~

> *Just for Blindsight's sake, in the case of a Big Bounce, it is interesting to note that the controversial Theosophist, H. P. Blavatsky, in 1888 alluded to something called the* **Breath of Brahman** *in her ponderous work,* **Secret Doctrine**~ *Oversimplified, she was referencing the Hindu tradition of denoting active periods in the Multiverse of expansion and contraction called Manvantara, also called Kalpas~ She wrote, "In sober truth, they are infinite; as they have never had a commencement, i.e., there never was a first Kalpa, nor will there ever be a last one, in Eternity."*

It seems the deeper we delve into the mysteries of the physical universe, the more enigmatic they become~ Scientists seem to be at times more driven by the excitement and competition of the chase~ The very concept of a chase demands the duality of a pursuer and whomever or whatever is being pursued in an endless, ever-advancing collapse of wave function after wave function~ While the role of the observer is extremely stimulating Intellergetically to a scientifically minded Energy Signature, the supreme Observer remains ever more elusive in the background indifferently observing Intellergy's observations~

This is the ultimate quantum squirrel cage~

~~~*~~~

At this point, we would like to suggest that the reader find and watch the film What the Bleep Do We Know, arguably the most prolific independent cult film in history~ Its combining of quantum mechanics, neuropsychology, and spirituality make it a great vehicle for Thojourn as it rides the cusp of Qen4 and Qen5 realities~

~~~*~~~

Blindsight would ask what scientists expect from finding the ultimate fundamental particle: the original force or the *Theory of Everything?* It is reminiscent, in a way, of the Alchemists of old in pursuit of a way to change lead into gold~ One could assume that great Qen4 wealth, recognition, and even power could be gained from the understanding and harnessing of this knowledge~

In the related fields of Biotechnology, the Terra Signature is experiencing a tremendous infusion of energy and information, with advancements such as stem cells and genetic engineering~

Cures for the likes of Cancer, AIDS, Parkinson's Disease, Alzheimer's, and many others appear to be just over the horizon~

However, at the same time, trillions of life-threatening microbes are on the advance, with increasing resistance to any and all efforts to slow, stop or eradicate them~

Seventy thousand years ago, tuberculosis emerged and has been called *history's biggest killer~* In the early 1900s, one in seven Americans and Europeans died of what was then called *consumption,* while thousands of others lived out their lives quarantined in hospitals and sanitariums~ Streptomycin came along in 1943 and became the first protocol proven effective against the disease~ The discovery of later antibiotics virtually eradicated TB, but it has begun a resurgence, with more than ten thousand cases reported in the US alone every year~ Of these, over a thousand do not respond well or at all to current antibiotics~

Staphylococcus aureus, contracted through medical procedures as simple as the insertion of a catheter, can cause a systemic infection~ It is extremely rare, but there is no known treatment or cure at this time~

Carbapenem-resistant Enterobacteriaceae (CRE), of which E. coli is a family member, is resistant to the strongest antibiotics we currently have, and about half of all who contract it will die~

These are just three of a growing army of microbes that many Light Bearers recognize as the Terra Signature's immune system reacting to the Human Disease infecting it~

Biotech has a plan, of course, with the likes of Phage Therapy, continued boosting of antibiotics and Quorum-sensing Inhibition, and these scientists are very confident and encouraged~ At the same time, microbes reproduce about every twenty minutes~ Each time, their genetic makeup learns a little something more about how to survive a little better, a little longer, a little stronger~ Humans, on the other hand, reproduce on an average of every twenty years, with a much slower adaptation rate~ Microbes have been on the earth for about 3.7 billion years, composing the most successful biology in the Terra Signature~ Sapiens have been here maybe one hundred thousand years~ How are we doing so far on the survival scale?

All this doom and gloom is not meant to instill fear, but rather a challenge~ Blindsight would offer that sustained higher frequency intention of intelligence and energy have all of the tools to resist and heal onslaught from any infection, bacterial or viral, on an energetic level~ There is an ample and compelling history of shamans, quantum healers, Reiki masters, and the like activating trained intention with amazing healing accounts~ The inconsistencies and failures over time seem to far outweigh the successes in these cases~ However, the mere fact that miraculously compelling reversals and remissions have occurred in many circumstances where all accepted medical protocols had failed begs questions that most likely still have answers~ From a quantum mechanical and Intellergetic perspective, perhaps we have simply not discovered the right questions~

At the same time, Blindsight is convinced that awareness as a fully entangled, self-aware Signome not only resolves and virtually erases all these questions and pursuits, but also opens into the infinite awareness and creativity of _____~

It becomes clearer and clearer that we are chasing that which very simply cannot be caught~

There will not be any mathematics in this piece and very few references~ There will be no grand explanations of Quantum Theory, as there is more than a century of documented sources available~

Anyone who claims to understand quantum theory
is either lying or crazy~
— Richard Feynman

With that as our premise, we invite you to join us for this limited, oversimplified examination of Blindsight, and we would remind you that if at any time you may feel you either understand *or* misunderstand any of what we posit, please remember we know absolutely nothing about what it is we write~

There are three Quantum concepts that we must elaborate on before we can truly continue, and those are **Entanglement, Superposition,** *and* **Time~**

Famalaich by Erin Nicole

VII

Entanglement

~~~*~~~

Who sees all beings in his own self
and his own self in all beings loses all fear~
— Isa Upanishad

~~~*~~~

ENTANGLEMENT is a proven quantum phenomenon embracing the idea that all perceived objects, thoughts, actions, forces, and phenomena were inextricably linked at the singularity of the Big Bang in terms of energy, intelligent information, and the intention of _____~

*(For later, there is a compelling discussion on **ancestor simulation** in our chapter, The Qen Trinity, relating the possible existence of some other intermediating, meta-intelligent source that may or may not exist between our Multiverse and _____~)*

Albert Einstein's Theory of Relativity insisted that nothing in the universe could travel faster than the speed of light~ However, in 1935, Einstein, along with Boris Podolsky and Nathan Rosen, published a paper describing phenomena that later became known as the *EPR Paradox,* which would involve the transmission of information at superluminal speeds~ There are some simplified explanations available for this theorem if one is curious to research it, but keep in mind this *is* a paradox, so fasten your Intellergetic seat belts~

There is also a principle that came later, in 1964, from physicist John Stewart Bell~ Bell's Theorem has been referred to as *the most profound discovery in science~* One of the most exciting components of Bell's Theorem in concert with the EPR Paradox and its implications for Blindsight is that, with its confirmation of superluminal speeds, it compels the existence of free will that could not exist in a deterministic, unthinking universe, which would run much like a machine~ The Multiverse rather behaves like a vast Qentinuum, where there is something that travels faster than light and that is intelligent information, the very source of choice and creativity~

_____, as the Source of this Multiverse, has injected free will and choice for its (and our) observation and entertainment~

Why ask a question if one already knows the answer?

Why make the journey if one has already arrived at the destination?

Let us imagine two *twin* particles are generated that have a counterclockwise spin~ One of the particles is then taken to the other end of the universe where, through a polarity process, it has its spin reversed to a clockwise motion~ Because of their entanglement, the other twin, with no lapse in time, instantly reverses to a clockwise spin~ The implication is astounding when such a communication at Einstein's speed-of-light transmissions — to include our radios, TVs, microwaves, and so forth—would take millions of years traveling at the speed of light~

Einstein called this *spooky action at a distance* and argued that Quantum Mechanics was incomplete~ He remarked later that, "God does not play dice with the universe."

Unfortunately for Albert in this case, quantum entanglement has been verified over and over with photons, electrons, neutrinos, certain molecules, and even small diamonds~

It had always been thought that entangled particles, even when parted, entangled from the same point in space~ However, in 2017, *Physical Review Letters* published a study wherein a research team created entangled pairs of photons that had emerged from different points in space~

Study co-author David Andrews, professor of chemistry from the University of East Anglia in Britain, stated, "Until now, it has been assumed that such paired photons come from the same location~ Now, the identification of a new delocalized mechanism shows that each photon pair can be emitted from spatially separated points, introducing a new positional uncertainty of a fundamental quantum origin."

Using the tried and true method of firing a photon into a crystal where it dies and two new entangled photons are born together in the same location, time, and space where the other perished, Andrews and his team found that entangled pairs can actually originate in separate locales within the crystal~

Andrews goes on, "The place of birth of the two new photons need not be co-located, because it's possible to connect them in the vacuum field, which is a standard facet of quantum theory~ Throughout our universe, there is a background of residual energy, which you can't normally tap—it's an energy associated with light when there are no photons present, called vacuum fluctuations."

These scientists appear to have recognized a new aspect of quantum uncertainty that previous theories had overlooked, revealing just a little more of the *fuzzy* characteristics of quantum physics~

It is another step closer for science to confirm by experimentation what Blindsight assumes as a fundamental principle~

As of this writing, a young Australian doctoral candidate in physics, Martin Ringbauer, along with his associate, Ognyan Oreshkov of Brussels, and others have found some *wiggle room* in Bell's Theorem by taking into account not only the physical aspects of the event, but also the *situational* features~ Oreshkov presents that measurements and observations are ultimately subjective, and Ringbauer adds that, "It lays waste to our conventional notions of cause and effect."

> *You can always draw a bigger box~*
> — Martin Ringbauer

The question has now become, instead of *WHAT* is it that is being observed, rather *WHO* is the Observer in the process?

WHO is the Intellergy in the entangled system?

WHO is outside performing the observation or measurement?

If we were to respond that it is our Energy Signature that is the *WHO* doing the measuring as the probability cloud is disrupted at the wave function collapse, then *WHO* is observing *our* observation?

For Blindsight, the significance of entanglement is that it gives the millennial mind/brain a way of embracing the oneness of all things and all beings~ In that spirit, Einstein was correct, because _____ does not engage in any dualistic game of *dice* with the Multiverse~

You're welcome, Albert~

NOTES

Reultach Air Sgeith by Kelci Auliya

VIII

Superposition

SUPERPOSITION is another very elusive concept that can challenge even the most fluid belief system~ The superposition principle says that an isolated quantum system or state can be depicted as a sum of two or more distinct systems or states~ More simply put, it means that one thing can exist in more than one place at the same time~ The reader will be able to find photographs of a single electron at near-light speed appearing in three or four different positions in the same picture~ Current experiments have confirmed superposition in lone neutrinos 450 miles apart~ The theorem, in concert with quantum entanglement, compels this effect even at infinite distances~

You might want to imagine yourself sitting across from yourself right now~

Another *spooky* aspect of superposition is the premise that net response at a certain time and place as a result of two or more stimuli is the sum total of the responses that would have been caused by each stimulus by themselves~

This would imply that if you were to yell, your other self would yell at the same time, and the overall volume of your shout would double as the totality of force~

The significance for communication and computing is staggering~ As of this writing, the *Quantum Computer* has become a looming reality~ Once believed to be decades away, it may be possible in less than ten years as of this simulation~ In a fully developed quantum computer, there are no zeroes or ones~ A one is effectively a zero and vice versa~ Information transfer, processing, storage, and transmission are not limited by the speed-of-light time lapse of the current digital world~ All processes are instantaneous, and communication with another quantum computer would also be instant, no matter where in the Multiverse it was located~ There is incredible hope and promise that such technology would bring with it the ability to solve problems that have long eluded current digital computation~ Those problems that would have previously taken extremely long periods of time would find solution in an instant~ Such a machine would be able to easily perform massive numbers of calculations, when a modern computer might require more time than the entire Multiverse has existed~ The possibilities for medicine, the environment, communication, all kinds of research, and virtually every aspect of the Terra Signature would become infinite in potential~

The reason we used the word *looming* is because, in certain models regarding the potential of a quantum computer, such a machine would theoretically be able to *absorb* the Internet~ Even the most complex algorithms of layered security from power grids to satellite communications, to air traffic control, to banking or even weapons systems would instantly become irrelevant, as though they never existed~ Unfortunately as a result, the competition to develop the first quantum computer has become the single most aggressive *arms race* in the history of the planet~ It is hard to imagine this type of technology in the hands of the wrong Intellergies~ It is also difficult to conceive who the right hands would even be~ As of now, factions ranging from the NSA in the United States to similar factions in Russia and China, all the way to Microsoft and Google, have collectively thrown trillions of dollars into this research and development~

At the same time, experts have told us Artificial Intelligence would not be possible until at least 2030, but in March 2016, Google's Deep Mind AI research company revealed a program they call AlphaGo which defeated 18-time world champion Go player Lee Sedol, 4-1, in South Korea in an historic competition~

Go is a very simple board game developed in ancient China; however, mastering this elegant simplicity is extremely difficult, as the legal moves on the 19-by-19 grid are very nearly infinite~ Google's AI team has worked for years to develop an algorithm with the complexity to challenge the human Go Masters~

AlphaGo continually *chooses* its next move by playing the rest of the game over and over in its *imagination* with two simultaneous artificial neural networks~ One network predicts the next move, while the other assesses the winner of each existing board position~

AlphaGo had analyzed about thirty million Go moves by human masters~ Then, by playing itself thousands of times, it began to discover new strategies~

In the ultimate face-off with Sedol, AlphaGo suddenly performed a move no professional player would ever make~ The *move* was so decisive, unexpected, and unorthodox, Sedol had to leave the room to regain his composure~

Is our very limited, primitive mastery of Quantum Superposition and Entanglement a blessing or a curse? We guess we will see~

After all this science, the story for Blindsight is quite different~ For example, in the rich Hindu tradition, there are many witnessed accounts of highly evolved spiritual masters seen walking and talking with friends and students while their physical body sat in meditation surrounded by followers many miles away~ Some of these masters were never able to be photographed and appeared absent even in group photographs~

Whether these reports are true or otherwise is not what is relevant from a Blindsight standpoint~ What *is* relevant is that quantum mechanics allows for these possibilities of out-of-body projection, mental telepathy, telekinesis, and other similar kinds of human metaphysical potentials previously dismissed as fantasy, occult, or science fiction~ Millions of Christians live in the confident belief that Jesus walked on water, turned water into wine, and brought Lazarus back to life~ Are these and other of his *works* any more extraordinary than those of the Hindu spiritual masters in a Multiverse where superposition and entanglement are present?

This is one odd, strange, unusual, and wonderfully weird Multiverse in which we find ourselves~ The mission, should you decide to accept it, is to fearlessly celebrate and playfully embrace it~

It's only weird if you're not~
— MWUSO

NOTES

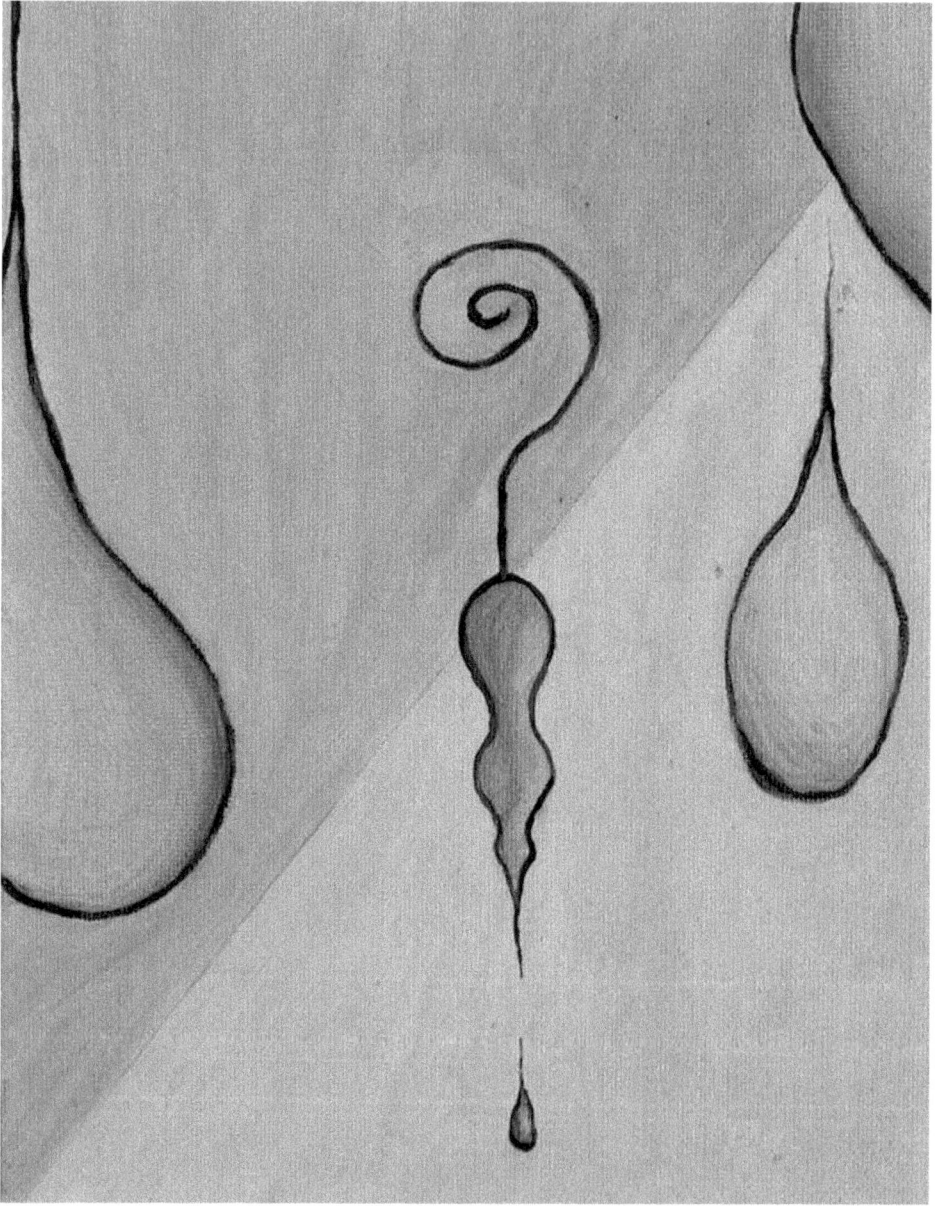

Uair by Erin Nicole

IX

Time

Time flies~
Time drags on~
It's about time~
Just in time~
Doing time~
Time is money~
Time after time~
Time stands still~
Time's up~

Time is on My Side
— The Rolling Stones

Time is the fire in which we burn~
— Delmore Schwartz

The only reason for time
is so everything doesn't happen at once~
— Albert Einstein

Apparently, time is ultimately a matter of opinion or, at least, circumstance~ Time is that mystical icing on the cake of the three spatial Qens that creates motion, memory, action, design, creativity, and potential~ It allows for the possibilities of measurement, observation, sequencing, and quantification~ It has the appearance of being irreversible, and it has the curiously wonderful characteristic that it is very difficult to be defined without using itself as part of the definition~ Try talking about time for very long without talking about time~

Intellergies such as Isaac Newton saw time as a fundamental structure of the universe~ From his point of view, it exists as a dimension unto itself, independent of actions and events, but where these actions and events occur in order~ Emmanuel Kant and Gottfried Leibnitz viewed time as a mind/brain configuration used by humans to view and sequence events~

It always seems to come down to that matter of opinion, doesn't it?

Is time felt as a sensation?

Is time more of a personal judgment call?

Just after the turn of the twentieth century, a virtually unknown physicist, Hermann Minkowski, inspired by the recent tenets of Special Relativity, conjured a mathematical structure called Minkowski Space, in which space and time became conceptually inseparable~ Spacetime was born~ Evidently, it was also determined to be curved and not linear as traditional geometry would have us believe~

In July 2016 at the "It from Qubit" meeting at the Perimeter Institute for Theoretical Physics in Ontario, several hundred physicists joined in to pursue Quantum Gravity, a concept based on the idea that space is actually made up of tiny pieces of energetic information~ By extension, they propose these chunks of Intellergy may be interacting with each other to create Spacetime~ Since the introduction of Quantum Mechanics, it has been one of the greatest challenges of physicists to blend with General Relativity, especially in terms of its relationship to gravity~ There is a theory that one of Spacetime's properties, which is its curvature, could be the fundamental cause of gravity~ For physicists, this concept holds promise of a combined Quantum/General Relativity Theory~

If Blindsight has to ask the physicists if they have already agreed there is no such thing as matter to be influenced by gravity, then why not ask the questions in a more Intellergetic fashion? It is simply further confirmation that science is moving closer and closer to Intellergetic understandings of the Multiverse~

There is a hypothetical topological bridge in Quantum Mechanics called the Einstein-Rosen Bridge; it is more popularly referred to as a Wormhole~ According to the theory, if one were able to master the ability to traverse a Wormhole, one could travel billions of light years instantly or simply go next door~ One could also theoretically enter other Qen strata or travel through time~

Beam us up, Scotty!
— Captain Kirk

These hypotheticals are great fodder for imagination, science, and fantasy fiction, but they are also fundamental tools for manifestation in a Qen4 Multiverse~

Fast forward to the early twenty-first century, when particle accelerators such as the Large Hadron Collider have produced certain effects that appear to occur before their causes~ Physicists generally do not speak of moving or traveling through time, but would rather refer to changes in spatial position with respect to the possibilities of closed timelike curves~

We are beginning to travel a little too far down the Wormhole here, so please continue this little side Thojourn on your own if so desired~

For Blindsight, time is a tool for Thojourn construction, but it is also a distraction, contributing to the illusory nature of Qen4 Intellergy~ As a result, we would rather focus on the absence of time and motion~ We are referring simply to NOW~

NOW is another indescribable concept that can only be experienced, and yet it is the singular most essential platform for self-actualization~ Please begin to breathe, live, play, love, and create from where you are, and WHEN you are~ Here follows a few Blindsight fingers pointing at the moon that is NOW:

> *...there is something essential about the now*
> *that is just outside the realm of science~*
> — Albert Einstein

The state of NOW is oneness with _____~ This is a state of beingness where there is no object, no subject, no motion; there is only infinite space, unlimited XPotential~ The experience of fully expanded space can appear uninteresting, even boring, to a Qen4 mind/brain driven by the turbulence-attracting densities of acceleration, interaction, and the tensions they manifest~ It is the realization of the XPotential factor that fuels the quest for higher Intellergy~

In reality, a fully entangled Intellergetic Signome is _____ with Nirvana awareness indescribable to a Qen4 Intellergy~ Once experienced, all intelligence from lower vibrational levels becomes shallow, limited, and illusory~ Fully expanded NOW space is the Source of all gods, Multiverses, and creation~

NOW is power~ The self-aware present moment is the singular opportunity an Energy Signature has to influence proactive change in the density of its Intellergy and create reality~

NOW is XPotential~ In the self-aware present moment, your next intention exists in a probability cloud of limitless possibility~ The next waveform that collapses has every opportunity to be one that is the result of proactive, mindful intention, rather than a reactive consequence~

NOW is grace~ In the spirit of *grace under fire* or *composure under pressure,* the self-aware, mindful moment makes possible the elimination of all anxieties or fears that make up the densities accumulated in a pressured, dangerous, or highly competitive instant~ It creates a relaxed and focused state of controlled, unperturbed intention for thought, observance, creativity, and action~

NOW is peace~ John C. Lilly referred to this self-aware, present space as the Center of the Cyclone~ One Mind at its Qen4 peripheral latitudes is very preoccupied with the process of creation, and the result is a turbulent, random, wonderful, chaotic, and beautiful multiversal cyclone~ Timelessness lives in the eye of the Qen4 storm, and that is NOW~ Without movement through time, the illusions of tension, duality, turbulence, and density cannot exist~

> *. . . be so utterly, so completely present that no problem,*
> *no suffering, nothing that is not who you are in your*
> *essence, can survive in you~ In the Now, in the absence*
> *of time, all your problems dissolve~ Suffering needs time;*
> *it cannot survive in the now~*
> — Eckhart Tolle

Zen is not attained by mirror-wiping mediation, but by self-forgetfulness in the existential present of life here and now~ We do not come, we are~ Don't strive to become, but be~
— Bruce Lee

NOW is NOW~ Time does not move or flow or stop or fly or have any true motion at all~ There is only the present moment~

If you miss the present moment, you miss your appointment with life~ That is very serious!
— Thich Nhat Hanh

One's experience of time should be viewed as a progression of reality in segments, not a flow of events~ If one were to throw you a ball, for example, your Intellergy would begin to react to velocity, trajectory, and position, based on prior accumulated and expected densities with respect to the process of pitch and catch~ If the ball is caught, the mind/brain will have had to receive the information, store it, process it, and transmit it to the consciousness for realization before you actually experience it~ All this takes place so quickly that the Energy Signature perceives it as a flow rather than a series of samples from the spectrum of reality~

How many times have you stubbed your toe, knowing that within a second or two you are going to feel the pain, while you wait as the information from the impact travels the neural highways to your brain and back? This is a classic NOW revelation as singular samples of the time spectrum are revealed~

When Intellergy is self-actualized in the NOW and mindful moment, there is no flow or motion~ There is only this instant and that is all there will ever be~ If there is a thought of a past experience, it has been brought into NOW for perception~ This experience is no longer, nor has it ever been in the past~ This happening has been transformed and filtered through all the accumulated densities that have been attracted as harmonics and resonances throughout the manifestations of the Energy Signature~

If one begins to plan for some future purpose, activity, or event, one starts by retrieving derivative and formulaic energy and information from accumulations in their Energy Signature's memory~ Just like a computer, our Intellergy is received, stored, processed, and transmitted, and so one engages one's own Intellergetic archives or database~ These acts all take place in the NOW~ All Intellergy must be brought into the present, processed in the present, and transmitted in the present~ If one reaches out to other sources, such as another Energy Signature or the Internet, all of this and future reactive events will occur only in the NOW~

At this time, please pick up your cell phone and retrieve a favorite photo of a loved one, a friend, a pet, or an event~ As you look at it, you are viewing a piece of NOW captured Intellergetically as ones and zeroes on your device~ Hopefully it is simpler to understand NOW if, by looking at the picture, you can imagine that three significant things are happening:

1) *You have just brought your past into the present~*

They say, "Your past catches up to you." It just did!

This is your simplest form of time travel~ The past does not exist in any way or form other than your observance and your interpretation of that memory in the NOW, filtered through accumulated densities, expectations, and current perceptions relative to present influences, forces, and intentions~

As you view the photo in the NOW, you most likely have very little Intellergy with regards to what day of the week it was, what the temperature was, or even what you were wearing when you snapped the shot~ Resonances and harmonics of the picture may be of a more fluid nature, such as happiness, love, or regret~ In fact, perceptions may be of any sense, but certainly you must understand they are very limited and selective relative to your own Energy Signature's density storage~ Just as a computer cannot process, interpret, or transmit information that it has not already received and stored, so your own Intellergy must only respond to the densities within the accumulations of its own Intellergy~

At the same time, no energy is ever lost, so the entire cumulative experience's Intellergy is stored as informational density in the Energy Signature~ It is just that most often the Intellergetic mechanism for retrieval is compromised by derivative harmonic resonances that manifest as preference, preconception, expectation, and discrimination~

2) *The manifestation of you holding the phone with the photograph, along with your entire reality, including your body, your position, your momentum, your consciousness, and your perception of the picture are only existing NOW~*

Please view the photo and see that as soon as the picture was taken, that XPotential reality sample ceased to exist~ The wave function collapsed~ Each being, object, thought, or action advanced into the probability cloud, creating an infinite progression of buckling waveforms and illusory realities~ Energy Signatures collect these types of densities in the perceived form of images, souvenir objects, or memories that in reality are no more than flickers of energy and information in perpetual oscillation~

3) *If you are looking at the photograph, thinking about it or planning what you could do about it in the so-called future, please remember you are doing all of that NOW~*

Time is NOW~
It is only NOW~
It can only ever be NOW~

Please live with that~
Live from that~
Live that~

> *I bought a cheap watch from a crazy man~ . . .*
> *It always just says now~*
> — Jimmy Buffett

"Breathe In, Breathe Out, Move On"

Should there be thought or Thojourn of a future or creatively visual event, it has also been brought into the NOW for perception and processing~ This occurrence certainly has no reality in an unforeseen future and yet too many times, anxieties and fears arise for non-present Intellergies to apply reactive intention that are often turbulent and self-destructive~

NOW is only NOW~ It is your most precious possession; ultimately, it is your only possession~

With intimate, mindful knowing of the present comes wisdom and the knowingness of _____~

Beatha by Erin Nicole

X

A Quantum First Impression

Would you like to hear a Quantum Impression?
Very well~
Here is an impression of our impression doing an impression of
  us~
Are you ready?
Here goes~
Would you like to hear a Quantum Impression?
Nailed it~

~~~*~~~

Welcome to a Multiverse where nothing matters~
This is a Multiverse where everything matters~

~~~*~~~

The Multiverse allows for any reality one will choose~
Do you want to be a hero? You got it~

Do you want to be racist? No problem~

Would you like to be a religious fanatic? Sure, why not?

Do you want to experience your reality as:

An Astronaut?

A Farmer?

A Soldier?

A Mother?

A Serial Killer?

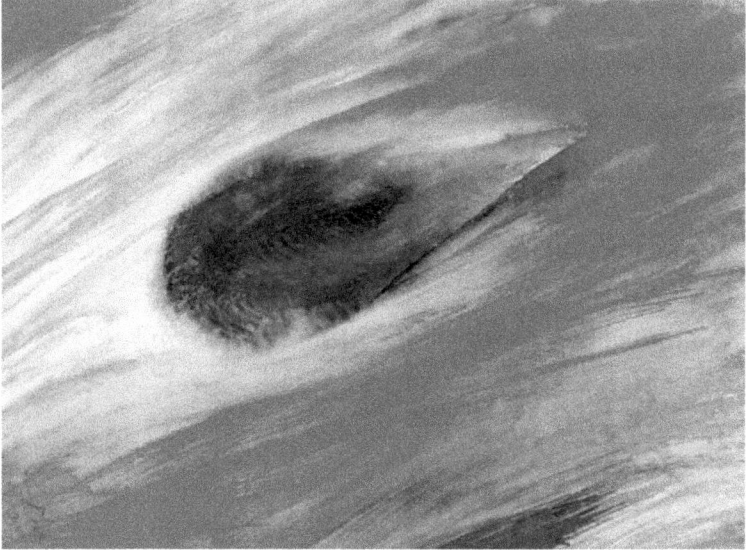

Fiorfe Gailleann by Erin Nicole

The Multiverse will respond to your intention inasmuch
as you are focused and committed~
Thoughts are magnetic and thoughts have a frequency~
As you think thoughts, they are sent out into the universe
and they magnetically attract all like things that are on
the same frequency~

Everything sent out returns to the source~
— Rhonda Byrne

Therefore I tell you, whatever you ask for in prayer,
Believe that you have received it, and it will be yours~
— Jesus

Everything is energy and that's all there is to it~
Match the frequency of the reality you want, and you
cannot help but get that reality~
This is not philosophy~ This is Physics~
— Bashar

Give it all your energy and let yourself believe,
for in this time, what you give is
what you will receive~
— The Tortugans

In truth, we really shouldn't be meeting here and now like this, because from a theoretical quantum perspective, the universe should not exist~

One fundamental premise of Quantum Mechanical Particle Physics states that equal amounts of matter and antimatter were created at the Big Bang~ Given the fact that matter and antimatter particles annihilate each other upon contact, how then could it be possible there are galaxies made up of trillions of stars with apparent mass? How can there be an appearance of mass of any kind? How can there be life forms and intelligent beings?

Theoretically, the cosmos should be a desolate wasteland of remaining radiant energy with no form~ The Big Bang actually should have resulted in an immediate Big Crunch~ Physicists in the millennium look to Dark Matter, Dark Energy, the Higgs Field, Neutrino Oscillation Anomalies, and the like on a seemingly never-ending quest to drill down to the most finite, microcosmic depths of creation, searching for the reason that we are here and knowing that we are~

It will be the Blindsight premise to contest that even should these brilliant scientific and mathematical minds discover this Holy Grail in the form of a fundamental theory, particle, or force, they still will not have found the Source~

Our simulated *book* is designed as a bridge~ The idea of a bridge implies a pathway leading from one side of an obstacle or depression while connecting one point to another~ Our Blindsight *bridge* is designed as an intelligent, energetic connection between one level or frequency of thought and another~

From our starting point on this side of the bridge, the terrain is made up of Qen4 reality: beings, objects, imagery, preconceptions, and belief systems that, in existence or expression, consist of height, width, depth, and are capable of movement through time~ These are the foundations of earthly human culture, philosophy, religion, institution, nationalism, education, and the rest~

From our bridge's substructure to its piers and spans, the Blindsight *bridge* must be constructed of these same Qen4 materials in the form of conceptual language~ By its very nature, language is inaccurate and misleading~ Nevertheless, words remain our only choice of conveyance for the possibility of reaching our destination on the other side~

It is the design of this bridge to make use of these modest and flawed tools of language and imagery to construct a conceptual roadway for those choosing to cross over~ A reader who would elect to make this crossing should understand that your *vehicle* is of Qen4 construction~ It is made of bone and tissue with a mind/brain *engine* of sorts fueled and powered by derivative, formulaic, and predictable intelligence~ With the introduction of multiversal *additives* of Intention offered along the journey, a traveler can learn to explore their reality and existence from a Qen5 perspective~

We hope your destination, like ours, is to connect with and enhance your existence as an Intellergetic being in a Quantum Fringe context, consisting of unlimited energy, mindful intelligence with infinite creativity, and self-actualized intention~

It is our goal to search out the wonder, elegance, and promise that abounds at the fringes of quantum mechanics and the farthest reaches of conscious presence~ It is our desire to attract readers with little or no technical background and a limited interest in the mathematics or the specifications of the science~ We are reaching for others like ourselves who are hungry to understand their true Intellergetic, metaphysical natures within a quantum millennial context blended into ancient consciousness exploration~

<p style="text-align:center">~~~*~~~</p>

Quantum physics came into existence in the early twentieth century through a magnificent explosion of Intellergy in the minds of the likes of Max Planck, Niels Bohr, Werner Heisenberg, Louis de Broglie, Arthur Compton, Albert Einstein, Erwin Schrodinger, Max Born, John von Neumann, Paul Dirac, Enrico Fermi, Wolfgang Pauli, Max van Laue, Freeman Dyson, David Hilbert, Wilhelm Wien, Satyendra Nath Bose, Arnold Sommerfeld, Nikola Tesla, and others~ What made this confluence so different from the historic emergence of great minds such as Galileo, Copernicus, Aristotle, Pythagoras, Da Vinci, and the rest was that these great thinkers all manifested at a time when communication between minds had been vastly accelerated as a result of the telegraph, the telephone, the steam engine, the automobile, and the airplane~ This communication *superhighway* of the times fostered a collaboration never before seen on earth~

Blindsight recognizes that this was the single most focused amount of collective Intention ever infused into the One Mind, which is always ready to manifest any reality, answer any question asked~ By asking questions of light quanta, or photons, they were able to observe and measure phenomena as particulate or wave-like~

Physicists had begun to recognize qualities in the fundamental properties of light, mass, energy, and force that completely overturned classical physics and the human views of reality~ They found it necessary to describe the results of their experiments on fundamental particles with almost mystical descriptions such as *strange, charm, flavors, WIMP,* or the *particle zoo~*

Depending on how a photon of light was observed, it could behave like a wave at times, and at others like a particle~ It was possible to observe a particle and determine its position, but one would not be able to calculate its momentum at the same time and vice versa~ For many of these scientists, the language of quantum mechanics began to take on nuances reflective of ancient philosophies like metaphysics, the paranormal, alchemy, or teachings of the Hindu and Laozi~

I'm very moved by Chaos Theory and that sense of
energy, that Quantum Physics~ We don't really, in
Hindu tradition, have a father figure of God~ It's about

Cosmic Energy, a little spark of which is inside every one
of us as the soul~
— Bharati Mukherjee

If Quantum Physics hasn't profoundly shocked you,
you haven't understood it yet~
— Niels Bohr

Brilliant mathematical constructs unfolded, revealing this newly discovered microcosm of reality~ Complex experiments and highly technical observational devices such as particle accelerators were soon to follow, continuing on into the millennium~ One Mind continues to answer all questions asked, reveal all observations sought~

Would this last hundred years have unfolded differently had this great intention movement focused its vast Intellergy on the Qen of intelligence and energy that preceded the manifestations they have pursued and observed?

Does the belief system of Quantum Mechanics hold any more observational relevance or significance for One Mind than that of a monk blissfully immersed in _____ high on a mountaintop in Tibet?

It seems worth noting that at the same time all these brilliant physicists were observing the Multiverse from an experimental point of intention, there was another eruption of Intellergy in the form of highly evolved Energy Signatures who observed the Multiverse from more of an experiential perspective; that is, from within their own minds~ Some of these included G. I. Gurdjieff, Helena Blavatsky, Rudolf Steiner, P. D. Ouspensky, Aurobindo, Henry Steel Olcott, Black Elk, William Quan Judge, Yogananda, Alfred Percy Sinnett, Aleister Crowley, Edgar Cayce, and Walter Yeeling Evans-Wentz, among others~

Through intentive meditation and contemplation, they were able to reveal manifestations in the Multiverse not apparent in laboratory experiments~ Most of them delved into metaphysics and spiritualism, while others explored the arcane, the occult, and Ectenic Force~ The personal realities they uncovered held as much value and worth for them as did the dual properties of light for Einstein and his associates~

In a Multiverse where there are an estimated three hundred sextillion stars amid two hundred billion galaxies, do either of these observational points of view weigh in with any more importance?

It seems fitting to quote Dr. Lilly once more~

In the province of connected minds, what the network believes to be true either is true or becomes true within certain limits to be found experientially or experimentally~ These limits are further beliefs to be transcended~ In the network's mind, there are no limits . . .

It is this fine blending of practical theorem and experiment with ancient philosophy and contemplative experience that has inspired the writing of this simulation~

For those readers who will bear with us through the maze of language found in most of these renderings, we promise a simpler, more loving and playful conclusion~

NOTES

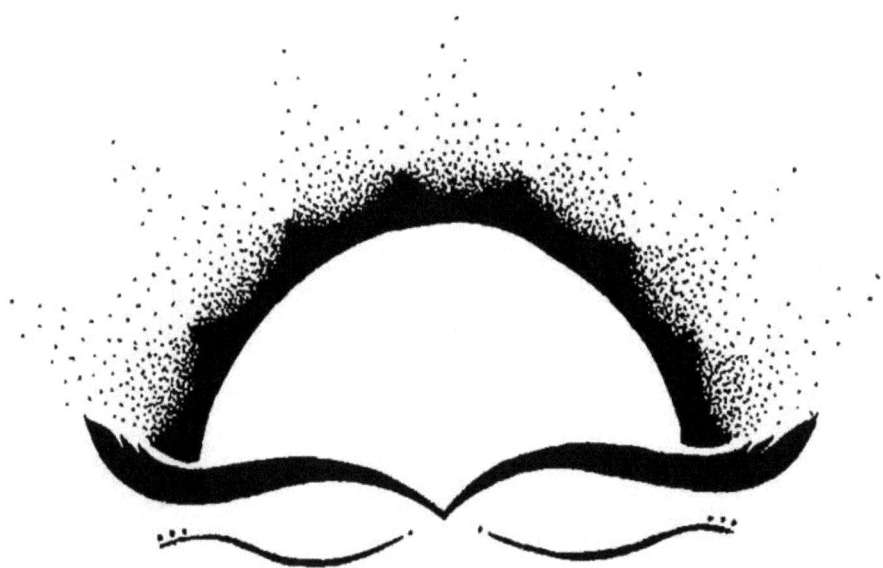

XI

Quantafiers

Following are some of what we call *Quantafiers* that we feel are important to form a basis for the possibilities of mutual understanding and communication~

Firstly, all language, quotations, or written words herein are to be regarded as limited and ultimately inadequate~ At no time will any statement by ourselves or any reference be looked upon as representative of a fixed reality or concept~ We are attempting to express Quantum Fringe concepts with extra-quantum language~

Given that language is our only method of conveyance, we make apologies for the absence of elegance with regards to descriptions of quantum ideas~ The Bible, the Bhagavad Gita, the Torah, the Koran, the Upanishads, Buddhist treatises and the like express remarkable beauty and grace, enabling study, institution, and belief~ As we begin to fashion simulations of entanglement, superposition, Multiverses, and the rest, it becomes readily apparent that very little elegance is conjured up in the average reader's imagination by these and other scientific terms~ It is out of our own struggles with the art of the language that new words and phrases have emerged in an effort to create different constructs for describing the indescribable~ It is also our hope that through the Thojourns we will embark upon below, readers will find their own elegance and bridges to understanding~

It is our desire to maintain as much simplicity amid the intricacy as possible in this narrative~ We will avoid expanded detail over references to personalities, phenomena, theories, experiments, or events that can be easily researched in the myriad sources available in the millennium~ Most of those drawn to these subjects are already familiar with basic, core principles and would find the repetition boring and tedious~

It is the singular goal of Blindsight to eliminate density from our Energy Signatures and facilitate immersion into the fully expanded space of _____~

We are energy~

We are intelligence infused with the One Mind~

Our self-awareness is our *signature,* just one way _____ observes itself and creates new Multiverses~

We are stardust~
We are Golden~
And we've got to get ourselves
back to the garden~
— Joni Mitchell

Often, we will use the words *play* or *playful* in any given approach, action, meditation, or visualization~ One of the fundamental premises of Blindsight is that One Mind is childlike, and our entanglement empowers our Intellergy to engage this aspect and enjoy our evolutionary path~

A body of clay, a mind full of play,
a moment's life –That's me~
— Harivansh Rai Bachchan

Man is most nearly himself when he achieves the
seriousness of a child at play~
— Heraclitus

For a human Energy Signature, play could be as simple and harmless as an infant with a stuffed animal, or it could be as intense as a 230-pound linebacker taking out a wide receiver on a crossing route~ Playfulness could be as calm and safe as floating on the surface of a pool on a warm summer day or diving off a ten-thousand-foot precipice wearing a wing suit~

For One Mind, play could appear as the elegant dance of a solar system in gravitational symmetry or the dynamism of a super nova~

Within every act or thought intentionally manifested as playful Intellergy, there are allowable densities accumulated that continue to resonate and attract further harmonics of play to the energy centers~ Densities with timbres tuned to energies of play are not only much more easily dissipated for transition into Qen5 space, but actually act as facilitators for the conscious evolutionary process by infusing flexibility, fluidity, and improvisational spontaneity into the Intellergy Forming of Thojourns that are created in construction of Qensional Bridges~

_____ infused unlimited XPotential for play into the Cosmic Inflation~ Lightheartedness and mischievousness are Blindsight's favorite toys~

Choose your toys well, do no harm, and play on~

Another Quantifier asks the reader to take a short time to ponder the very basic concept that everything in our existence begins with thought~ This would include every action, word, observation, or creation~ From the simplest functions such as breathing, there is an internal reactive movement or a trigger of brain impulse that precedes it~ There was never a building constructed, a song written, a discovery made, a relationship begun, or anything in our experience that did not begin in thought~ An estimated 99.999999 percent of all thought is reactive~ It is a reaction to intelligent energies and the self-aware intention of _____ set in motion by the Big Bang~ However, in the face of these seemingly irresistible forces, we would humbly suggest in the renderings to come that there is a *path* open for our sentience, which we call our Energy Signature, that allows us to become consciously and intentionally absorbed into _____, enabling infinite proactive creation of our reality instead of a limited reactive response existence.

NOTES

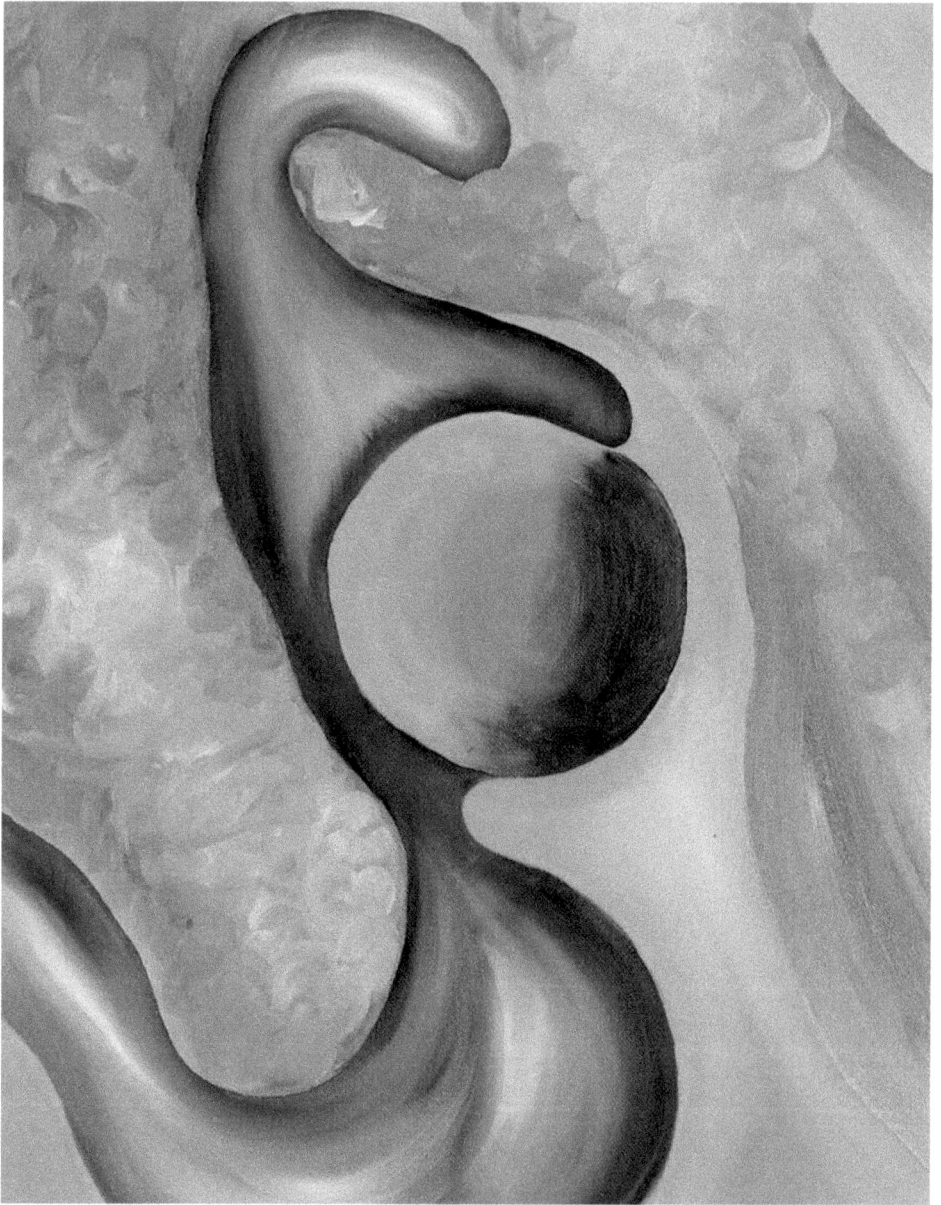

Tu'r by Erin Nicole

XII

Intention

Watch your thoughts, they become words~
Watch your words, they become actions~
Watch your actions, they become habits~
Watch your habits, they become character~
Watch your character, it becomes your destiny~
— Laozi

Intention appears to be something akin to a tuning fork,
causing the tuning forks of other
things in the universe to resonate at the same frequency~
— Lynne McTaggart

~~~*~~~

At this moment, simply hold your breath playfully for ten seconds and calmly observe~ You have just applied intention, altered your entire Intellergy in the mindful *NOW*, observed the event, collapsed a wave function, and transmitted new, Intellergy Formed impressions into One Mind and _____~

Well done!

The Intention of which we speak is deep, powerful, and heartfelt~ It resonates within the profound core of our being, and it requires committed mindfulness for its engagement and success~

In truth, intentive play with one's breathing is the most readily available, intimate beginning of a path to understanding the wonderful toy that is the power of intention~

Live your dream~
Create your life~
Lead with intention~
— Leslie Schwartz

God is a verb, not a noun~
— R. Buckminster Fuller

This chapter on Intention is placed *intentionally* here, early in our simulation, to set a tone for the evolution of our mutual understanding~ Our discussion of Intention is not only a reflection of our own mindfulness in the expression of this exercise, but also your own as a reader following these renderings~

Intention is the driving impetus of all creation from before the Big Bang with the intentive creation of _____ to the next wave function buckled by our next thought, word, deed, or observation~

We Energy Signatures use Intention unintentionally with every waking moment~ These *unintentions* are virtually all reactive, stemming from derivative accumulations of densities in the Intellergy sphere~ Within the simplest of actions, such as picking up a coffee cup or driving to the store, to the most complex such as brain surgery or unilateral disarmament talks, streams of unintention after unintention unfold Intellergy Forming reality perceptions in *NOW* segments~ These segments are fashioned of Qen4 density resonances from buildups in harmonics generated by desire and past experience related to like-dislike, fight-flight, good-bad, or whatever duality du jour happens to manifest~

A mindful, self-aware Intellergetic Being with even the slightest Qen5 contact experience is capable of stepping outside of this reactive flow of intention~ Like the neutrino that passes unimpaired and unaffected through the earth, creative choices are XPotentially made available to an Intellergetic Being who is unattached to reactive densities~

> *What happens is not as important*
> *as how you react to what happens~*
> — Thaddeus Golas

We have a friend who has a self-proclaimed bad temper~ They have seemingly come to enjoy the adrenaline rush of excitement that comes from its eruption~ Although we have never seen them act out violently, nevertheless acting out in angry ways has become a sort of addiction for them~ We have even told them that we purposely *light them up* just for fun with mention of a subject we know is heavily charged for them~ This offers both of us opportunities for observation, and usually a good laugh afterwards~ For their part, they are beginning to see how much more is to be gained in terms of their choices if they can learn to just stop, pause, and take a breath before launching into reactive Intellergy~

Mindfulness is the state from which self-aware intention is realized~ Each time we *react without thinking,* we activate density centers in the Energy Signature charged with the resonant harmonics of past experience~ Often these reactions would collapse entirely different, unique, and life-altering waveforms if driven by present and mindful intention~

> *Mindfulness is the observing things as they are.*
> *. . . without laying or adding any of our projections*
> *or expectations onto what is happening~*
> — Frank Jude Boccio

For us, such situations are the opportunity to be the neutrino, allowing these inflamed densities to pass through us observed, without reaction~ It is in these potentially intense encounters with powerful and concentrated Intellergy that *NOW* presence and awareness are incredibly enhanced, adding fuel to the engines of our transition~

Simple, mindful *NOW* intention can be very simple, with acts such as the holding of the breath~ One can begin with simple attentive and intentive acts, such as buckling a belt or putting a key in a lock~ If you are mindful, focused, and intentive, even the simplest of acts can take on a much larger, expanded context, stimulating conscious, evolutionary progression~

> *Attention is never free~*
> *This is why we pay~*
> — MWUSO

There is the story of the initiate in the Zen monastery who was struck across the back for brushing his teeth and peeing at the same time~ The elder monk who caned him was reprimanding the initiate for dividing his intention, his focus, and his oneness with the moment~

Difficulty inevitably arises when maintaining mindful intention in more complex or difficult endeavors~

> *If you can keep your head when all about you are losing*
> *theirs . . .*
> *If you can think and not make thoughts your aim . . .*
> *If you can wait and not be tired by waiting . . .*
> *Yours is the Earth*
> *and everything that's in it~*
> — Rudyard Kipling, "If"

These are extracts from one of our favorite poems, and it stands as a fine anthem for mindful intention and will~ We hope you will find, read, and reread the entire poem at some point~

The message is that Intention is simple in its initial phase, but it requires commitment and will for its Intellergy to build in more formidable endeavors to engender Intellergy Forming and the altering of derivative reality densities~

Gurdjieff, in addition to all of his vast interests and undertakings, had a fascination with the musical octave~ He observed that in the context of a vibrational Multiverse, harmonics, resonances, and timbres in an Intellergetic sense arise from various sources and directions constantly colliding, interfering, getting stronger or weaker, and so on~ He expounded that from the finest to the coarsest, these frequency modulations make up the entire Multiverse~ String theorists love this perspective, as they have found vibrational correlation in phenomena of heat, magnetism, light, chemical, and other reactions~

The seven-tone octave held particular interest for Gurdjieff with respect to intention, action, and creativity~ Through his extensive travel and research, he had determined this octave originated in a formula of *cosmic law* worked out by ancient arcane schools and later applied to music~

If we were to view the octave in the context of Do, Re, Mi, Fa, So, La, and Si, each step can be observed as an increase of vibrational frequency initiated by intention~ If observed musically as semitones, the intervals between Do and Re, Re and Mi consist of two semitones~ Between Mi and Fa, there is a change in vibration with only one semitone, like a retardation or a change in vibratory direction~ The next step from Fa to So then requires new intentive Intellergy to reestablish a two-semitone progression toward the completion of the octave~ This new intention must be so much more focused and powerful than the first Do that it carries the progression from So through La to Si~ Transition to the next higher octave with the manifestation of a newly created Do from Gurdjieff's perspective requires meta-human will, focus, and intention~

Take a moment and use this Gurdjieffian Thojourn from our experience to perhaps apply to a situation in your own manifestation~

~~~*~~~

We recently decided to clear all the brush and small trees that continually advance to take over our little ranch~ In some places, this spread of foliage was three or four feet inside the fence, which had become all but invisible most of the way around the property~

Do *– We mindfully created the intention to clear all the trees and shrubs from the fence line~*

Re *– We attached the brush cutter to our trimmer, fueled the machine, dressed for the project, and ventured out~*

Mi – *We began to trim and cut, with a project that would take several days to complete~*

Fa – *And then it began to rain~*

So – *We renewed our intention, refusing to allow the rain to stop us~*

La – *We continued to cut and trim, completely soaked, our glasses constantly fogged and dripping~*

Si – *The rain turned into a torrential downpour, driving us back into the barn~*

The vibrational **Do** *that would have represented the completion of our project was never realized that day~ It required a complete renewal of commitment and intention to return, generate a new octave, and complete the task on another day~*

Another day, a new **Do** *and yes, the job was done~*

~~~*~~~

Do is, for this Thojourn, much like the initial pluck of the string on our Multiversal Guitar~ The intention of this original intentive pluck was to trim a few trees, but what if the intention was to get a college education or take a trip around the world? What level of mindful, committed purpose would be required to muster the time and resources to complete octave after octave in the pursuit of such complex endeavors?

The act of conscious, present intention is quite simple with a little practice~ It is the continued focus and duration that is the challenge~

Intention is your toy for playfully forming realities in your Energy Signature, the Qen of the *Now* progression into your future~

Intention is your tool for identifying, excavating, and breaking up densities in your Intellergy~

Intention and density awareness are your weapons of defense against the Human Disease and Intellergy *Criminals* that would invade and attempt to introduce dissonant, strident harmonics to divert your octave progressions for evolution and transition to expanded space~

Meditate, ponder, and contemplate often the role Intention plays in your manifestation; its role as enabler in the Human Disease; its part in the creation of your very reality, your perception of self, and its impact on your potential for the future~

Imagine the octaves of your life played with *NOW* mindfulness, self-actualized consciousness, and fearless, playful Intention by an Intellergetic Being manifesting as its true nature of energy and intelligence~

Finally, remember the ultimate simplicity is _____~

MWUSO was quoted, "There is a point when all must be lost in order to find what might be gained." Immersion into _____ is the simple allowing of all density attachments of identity, possession, relationship, and duality to dissolve into fully expanded Intellergetic space~

NOTES

Blathan Iongantas by Kelci Auliya

XIII

Density

The concept of *DENSITY* is quite possibly the most important concept Blindsight attempts to express~ It is another term from the Expanded Glossary, offered for the transition seeker, that has been altered by Blindsight to allow for broader interpretation~ If there is one single idea the reader can take from this Thojourn, it should be the beginnings of a committed effort to recognize and dissolve all density in their Energy Signature~ Mindful Intention is the singular *tool* for the dissipation of Density~

There will be many references throughout to *density* and *densities* with respect to Energy Signatures, as well as larger perceptions and realities~ For a better understanding of the Blindsight use of these terms, we would challenge the reader once again to begin immediately playing with the notion that they and everything around them are made up of energy and information (or intelligence)~ The Multiverse is not solid or material in any way~ It only appears to be so because of the reactions and interactions with density awareness in the lower frequencies of our Intellergy fields~ This includes your body, the book you hold, the seat you are on, your surroundings, and ultimately everything of which you can conceive~

*If you want to find the secrets of the universe, think in
terms of energy, frequency, and vibration~*
— Nikola Tesla

*The energy of the mind
is the essence of life~*
— Aristotle

*Please take some time (and do it often) to ponder and
meditate on these ideas~*

Even though our Qen4 dimensional perspective appears
to separate our individual Energy Signature from everything
else, this is pure illusion~ We and everything in the Multiverse
are entangled in Quantum Intellergetic ways~ In the Quantafiers section, we asked the reader to imagine themselves as an
energetic body, and the implication at the time was that the
entity was somehow independent of its energy environment~
This exercise was offered in an introductory spirit, as most all
humans tend to perceive of themselves as unique individuals
in this dualistic fashion~

A more advanced vision of our true nature is to imagine
our perceived individual Energy Signature more as a temporary event immersed and in participation with the Spacetime
Continuum~ Much like a wave on the ocean that amasses energy, form, and motion only to be absorbed back into the whole,
or a spark from a fire that glows brightly and soars into the
night until its inevitable immersion into the Qen Intellergy
pool, our true nature is entangled and infused; we are infinite
in our intelligence and energy~

Sgalpaid by Kelci Auliya

Originating in Chaos Theory, there is a term called the *Butterfly Effect,* the name being derived from an early metaphor in the theory related to weather~ The example was given that the gentle flapping of the wings of a butterfly in China could influence a hurricane on the other side of the planet~ Whether we actualize the path of our awareness as conscious, expanded space or not, each and every thought, word, or deed sends ripples of Intellergy out into the Multiverse and provides the Universal Mind and _____ with a new aperture for observation~

From this perspective, what would be the possibilities of your potential? How fluid, malleable, and creative could your abilities evolve to influence and mold your reality? What kind of personal power could you wield as an unencumbered, fully entangled Intellergetic being instead of a dense, fragile meat suit?

Imagine that Thojourn involving a body/mind as that swirling probability field we discussed earlier~ For a moment, attempt to step outside of it and become an Observer of its existence~ We would see places in the field that may appear *darker* or not moving as fluidly~ These would be the *densities* that Blindsight describes~ These densities are collections, convergences, and accumulations of reactive Intellergy from past experiences, reinforced by expectation throughout the Intellergy field~ Densities have an appearance of mass that attracts other resonant or harmonic energies and information of like concentration or frequency~ These densities also cause Intellergy Forming in the Terra Signature with manifestations of religion, personality, money, sexuality, language, nationality, and all agreed-upon realities and belief systems in which humans participate~ Compacted Intellergy becomes acceptable, unquestioned certainty~ A religious or nationalistic belief system for some can be as solid a reality as the floor upon which they stand~

When you meet an interesting person, your Energy Signature already contains certain *density harmonics* of Intellergy relative to your experiences and your perception of body type, gender, intelligence, or personality~ The densities in your field immediately begin to resonate with those energetic traits in that person's energy sphere most conducive to furthering the exploration of the relationship~ At the same time, your densities are resonating with those attributes in the other sphere that may be clashing with some of your vibratory energy~ These could be things like perhaps they talk too much, which should be perceived not as talking, but rather a higher energetic output at a certain resonant frequency~

Instantly, the densities in your energy field begin to absorb, store, process, and then transmit reactions that stimulate further attractions and responses in a continuing cycle~ This cycle can repel fairly quickly or it could develop to create more resonant attractions leading to a possible progressive, interactive connection~

Another way density is accumulated in an Energy Signature is in reaction to a trauma or frightening circumstance~ For example, one may have had a bad experience early in their manifestation having been bitten by a snake~ In that moment, densities within the Energy Signature began to form respective to the pain, distress, and anxiety~ Densities also form from visual impressions such as the wound, the snake itself, or the surroundings~ Many develop an almost crippling fear of snakes from such an experience~ The simple thought of a snake, seeing one in a zoo, in a movie, or even finding themselves in a similar environment where a snake could live can evoke and enhance the resonating densities in their Intellergy field, attracting more and more density, or fear~

For an evolving Intellergetic Being, the observation of these reactive densities creates amazing opportunities for dissipation of inharmonious resonances~ In Qen4 terms, it is facing one's fear and choosing to make mindful *NOW* evolutionary decisions~ Seeing a snake in a zoo or finding oneself standing in high grass does not necessarily place one in danger~ More likely, the XPotential of *NOW* will enhance calm awareness and reaction time if needed~ Intellergy in such a state will literally melt away associative densities, creating harmonics much less likely to resonate as strongly in future encounters~

Many is the time one's Energy Signature comes into contact with another Intellergetic being that has a difference of opinion on a given subject~ These differences may manifest as either words of a conceptual nature or as actions that one or the other finds unacceptable~ A common case in point might be driving in traffic and finding oneself suddenly cut off by another vehicle moving into one's lane~ Many Intellergies have large amounts of density accumulated with such experiences~ As a result, these energy centers immediately begin to resonate with dissonant harmonics, manifesting in the form of their own negative language and gesture, while at the same time beginning cycles of destructive attraction and transmission~ In extreme cases, this has been known to accelerate into what has come to be called *road rage,* leading to confrontations and, at times, violence~

If, in this situation, an Energy Signature has done the difficult work of dissolving this type of Intellergy within its sphere, all of the intention and density of the event passes through without attachment~ A state of XPotentiality now exists to optimize reaction time, enrich awareness of surroundings, and offer the best opportunity for a safe outcome without any remaining resonance~ The absorption and recycling of such negative intentions merely reinforces existing density centers within the Energy Signature, ensuring the attraction of other such events~

~~~*~~~

There is a curious little particle in the Multiverse known as the *neutrino,* and it is Blindsight's personal favorite because it passes unimpeded and undetected through basic matter~ Since it has no electric charge and is essentially neutral, the electromagnetic force also has no effect on it~ Essentially, density has no impact on its intention, and that is why it is a tremendous basis for Thojourns~

From one perspective, an Energy Signature might creatively imagine itself as a neutrino passing through its manifestation unaffected by belief systems and formulaic thought~ A being in this state of *high indifference,* to quote Franklin Merrell-Wolff, could move and act within its manifestation while remaining in XPotentiality at all times, impervious to densities and accumulated harmonics~ Such a being would exist in a state of clarity and expanded awareness, literally on the cusp of Qen4 and Qen5 Intellergy~

From another point of view, this same Energy Signature would experience all intentive densities and dissonant harmonics passing through them with absolutely no effect or interaction~ This position affords the Intellergy the feeling of unconditional love and unbounded joy with no tension, judgment, or duality~ Imagine as we proceed how different the scenarios presented below would unfold infused with this view~

~~~*~~~

There was a time for us spent at the bottom of Palo Duro Canyon~ One day, hiking in the arroyos, we came upon a split rail corral where a remuda of horses was kept for trail rides through the canyon~ An old cowboy stood hipshot at one end with one foot up on the rail, watching a young wrangler, probably in his early twenties, trying to train a new gelding to a lunge line~

We walked up and leaned on the rail next to the old cowboy~ He looked over and smiled, tipping his slouch-brimmed hat, then he leaned over and spit a large brown wad of chaw over the fence~

"Name's Byron, but folks call me Dusty," he said and reached a gnarly, sunburned hand out to me~

We took his hand, gave our name, and turned back to the drama unfolding in the corral~

The young wrangler was having quite a difficult time with the horse that was rearing, eyes wide with fear~ The horse was having none of the wrangler~ We could see the cowpoke getting more and more frustrated until finally he reached over, grabbed a whip, and began hitting the horse~

Like a mountain cat, Dusty was suddenly over the fence and striding into the fray with a look on his grizzled face beyond anger~ He grabbed the lunge line, then the whip, and gave the young wrangler a crisp taste of it across his backside~

We couldn't quite make out what was said next, but it was loud and it was hard, because the cowpoke turned away fast, left the corral, and headed for outa sight~

Within less than a minute, Dusty had calmed the gelding and was leading him over to his stall for a fresh bucket of feed~

When he returned to the fence, he was shaking his head as he wiped it with his old blue bandana~

"I've got to say, Dusty" we said, "That was really something out there."

"Yeah," he replied, "I ain't sure the kid's got the makin's of a true wrangler~ He gets worked up a little too easy and them ponies, they pick up on that."

"I know horses are really smart, and they can be really hardheaded; what's your secret for keeping them all in line?" we asked~

"You gotta love 'em," he said, pushing his sweat-stained hat back on his head with a slight wrinkling in the corners of his eyes, "or they'll make you mad."

~~~*~~~

We had found one of the greatest lessons of our life in the presence of an old cowboy and a split rail corral at the bottom of a big red canyon~

The mindful Intention of Love is the path to engaging the neutrino Intellergy in our Energy Signature~ From that moment, we have tried to remember that anytime a difficult person or situation arises, just let the density pass right through unattached~

"You gotta love 'em, or they'll make you mad."

~~~*~~~

At the time of this writing, there are significant densities accumulating at the time of the Great Shift in the Terra Energy Signature with regards to race, politics, and religion~ Centuries of amassing extremely dense, reactive energies and information has resulted in compacted realities with almost no fluidity or flow~ Volatile reactions are brought about with inflammatory consequences~

The energy rift between Islam and Christianity is often believed to have begun with the Crusades, when British knights more than 400 years ago stormed into Jerusalem under the pretext of protecting the Holy Land and defending Christians in non-Christian lands~ Too many cases of plunder, rape, and slaughter of innocents inflamed Muslims and started a domino effect of reprisal after reprisal still in play centuries later~ With each act of revenge, the Intellergetic density of each belief system is compressed more and more, with little resemblance to the expanded origins of love and peace in the philosophies that fostered them~

Islam itself has many internal densities in the form of divisions such as Sunni, Shi'a, and Sufi, along with dozens of fundamental extremist factions~ The World Christian Encyclopedia estimates there are more than thirty-three thousand Christian denominations worldwide, with many of these devotees holding the belief that their proprietary understanding of their relationship to Jesus and their interpretations of biblical scripture is the only true path, leaving all other Christian faiths destined for damnation~ This absurd exclusivity creates further densities within those particular belief systems in ever-spiraling reactions~

Elimination, or at least reduction, of Density in one's own Energy Signature should be the main focus of the work from a Blindsight perspective~ Using the Thojourns offered in this book and other related works, one's unfaltering, dedicated intention and a spirit of play, the densities will begin to dissipate, allowing for more and more multiversal Intellergy immersion and flow~

Blindsight encourages you to begin immediately practicing interactive observation with all aspects of your reality and your perceptions of it in a playful, energetic manner~ The impact on your awareness and your ability to take ownership of creating your future will be immediate and life-altering~ The successes will come in short bursts at first, but these glimpses will create their own attracting natures, fostering more of the same with longer duration~

Playfulness will be elaborated more later, but it is important to understand at this juncture that as one begins to interact playfully with higher intention, expanded energy, and intelligence, the universe becomes more pliable and fluid~ One Mind literally loves to play~ Taking a loving and playful approach to one's transition toward Qen5 perception stimulates and attracts the same forces in the Multiverse~ Love and play accelerate evolution, creativity, and positive harmonics in the probability field~

> *Creativity is intelligence having fun~*
> — Albert Einstein

It is important to note, however, that at the same time, the universe loves and even rewards flow and consistency, even though it is not necessarily desirable from a Blindsight standpoint~

> *It's not what we do once in a while that shapes our lives~*
> *It's what we do consistently~*
> — Tony Robbins

> *The habits that took years to build,*
> *do not take a day to change~*
> — Susan Powter

> *Consistency is the last refuge of the unimaginative~*
> — Oscar Wilde

In Blindsight, consistency can become a detriment to achieving more expanded Intellergy~ Regularity and dependability are extremely valuable qualities leading to successful endeavor, skill, profit, and power in Qen4 reality, but continual attraction to repetitive Qen4 intention creates expectation, density, and formulaic predictability~ These assets continually accumulate Intellergetically Formed densities in the Energy Signature~

The Multiverse responds to all efforts of an Intellergetic Being evolving and expanding outside of their habitual flow with turbulence~ Multiversal forces are called into play that alter density structures, as they begin breaking apart and re-orienting their new attractive natures and resonances~ The introduction of new energy and intelligence stimulate at once very positive manifestations and at the same time some possibly turbulent results~ Examples of common, yet significant, human transitions might be changing careers, moving to another country where you do not speak the language, or ending a long-term relationship~ Fundamental shifts such as these immediately send ripples of energy and information into the probability field, collapsing wave functions and altering the Energy Signature in intimately essential ways, most often with completely unforeseen results~

In these endeavors, Blindsight would have the Thojourner celebrate the journey and allow the playfulness of the Multiverse to respond and propel your expansion~ There is a certain trust factor that emerges and grows as an Energy Signature gains confidence in their abilities at play in the One Mind~ Once an Intellergetic Being is able to know with sureness that their true nature is that of intelligence and energy, personal reality takes on more expanded and fluid tendencies~

> *Leap boldly off the precipice~*
> *Trust there will be a ledge~*
> — MWUSO

We have found driving a car to be a wonderful exercise in Intellergetic awareness~ For us, the vibration of the engine, the feeling of motion, and the constantly changing scenery facilitates harmonic states~ We use a combination *Motion Meditation* that involves taking regular, full deep breaths and holding an attitude of play at the forefront of our thoughts~ Keeping a slight smile in the eyes seems to engender presence of mind and a loving perspective~

In Buddhism, *walking meditation* is one of the most widespread practices~ Gautama pointed out five benefits of this method:

1) One is fit for long journeys~
2) One is fit for striving~
3) One has little disease~
4) That which is eaten, drunk, chewed, and tasted goes through proper digestion~
5) The composure attained by walking up and down is long lasting~

We would like to add a few more benefits from our own experience with *Motion Meditation* qualified with the addition of play:

6) There is an enhanced richness to every action and experience, whether positive or negative~
7) After long periods of this exercise, one will often laugh for absolutely no reason other than pure, uninhibited joy~

Sometimes your joy is the source of your smile,
but sometimes your smile can be the source of your joy~
— Thich Nhat Hanh

8) When in the state of playful, active meditation, one soon finds it to be contagious with whom one comes in contact~ Others you meet will often smile themselves more readily for no apparent reason~ Some actually

become more playful in the moment, stimulating a more fluid, improvisational, Intellergetic exchange~

9) From a health standpoint, smiles and playfulness free neuropeptides in the brain, warding off stress~ The *feel good* neurotransmitters of dopamine and serotonin are released, relaxing the body, lowering heart rate and blood pressure~ These also act as natural pain relievers~

10) We look better and younger when we smile and live in a state of play~ At the same time, we *feel* like we look better and younger!

Any activity can become meditative if deep breathing, a clear mind, and present, playful intention are engaged~ This could include eating, sports, working, or even making love~ It is an opportunity to observe the Intellergetic body in motion~ Sensations and perceptions are also more easily observed and enjoyed~

The millennial world has become so accelerated that few Energy Signatures can find the time or the attention span to simply and quietly sit in calm reflection~ *Motion Meditation* is a practical method for becoming more aware of one's Intellergy, remaining more at ease while in motion, and recovering more quickly from the tensions of active repetition~

Over time, your Energy Signature will begin to dissipate the densities within the field~ Think of this process like clearing dirt and rock debris from a stream to allow for better flow~ Indeed, there is an infinitely powerful flow of energy and intelligence from the Universal Mind attempting to move unencumbered through your field at all times~ This is the source of unlimited XPotential, intelligence, and power to create the design of your future~

Strive every day to make the fundamental understanding of the concept of *DENSITY* integral to your evolution~

Please don't misunderstand that Blindsight concepts and practices will dissolve at once all sensation of physicality, mass, or gravity in your existence~ What is offered is a glimpse into the possibilities that await those who would do the extremely hard work to become a fully realized and entangled Intellergetic being~

Dedicated intention applied to dissipation of density in your Energy Signature is ultimately your only goal~ In the rare and precious moments that all density is gone, you will find yourself in fully expanded space, immersed into _____~

Astar by Erin Nicole

XIV

Qen

Qen is ultimately indefinable, and yet a term must be conjured here for the expansion purposes of this simulation~ Qen is the infinite field, pool, continuum, medium from which all Multiverses emerge~ Qen is the undisturbed *palette* from which _____ mixes with intention the many *colors* of creation~

Qen may be Intellergetically experienced as a spatial aspect when observed as such, but Qen may also be perceived as particulate, with characteristics simulating mass~ Qen for this Multiverse exists as a Qentinuum within which the Multiverse evolves~ This is to include all manifestation, Intellergy, and Spacetime~

The closest analogy is the Quantum *dimension;* however, all dimensions evolve within Qen~ Qen assumes an infinite spectrum of frequency expansion when spanned across a Multiverse~ This spectrum has vibratory responses reflecting *strata* of harmonics and resonances like an FM radio station at 101.5 MHz that samples from the FM radio band between 88 and 108 MHz~ These strata, although completely entangled with the full spectrum, resonate with *bands* of frequencies that take on multiversal properties loosely individual to the respective sample of the spectrum~ Much like the radio station, these bands assume densities for perceptive resonance that become Intellergetic in structure~ Qen, in this context, may be assigned harmonics such as Qen4 or Qen5, affording an Energy Signature immersion access and/or experience for transition movement through Thojourn, meditation, hallucinogen, or some other form of mindful Intention~

By this time, it is assumed that, at least for this simulation, you have managed to displace a sufficient density of belief systems to mount a fundamental possibility that you are an Energy Signature made up of energy and intelligence~ You are in no way made of matter possessing mass or weight~

As an Intellergetic being, you are entangled with all other manifestations of Intellergy in all Multiverses, One Mind, and as such with _____~

All duality is illusory~

There is no path to be taken~

There is no destination at which to arrive~

You are always immersed in your true nature as infinite, expanded, consciously energetic space~

All that is required is simple, revealed, loving acceptance~

From this *revealing,* intention is what creates your perceptions of your Multiverse~

Qen is our truth~

Blindsight is not, nor could it ever be, your truth, your clarity~

Qen exists for you as you perceive the densities and attractive forces of your Energy Signature~ Blindsight as a simulation is a duality; something to be read, absorbed, processed, and transmitted, usually somewhere between acceptance and rejection~

Qen is merely a *finger pointing at the moon* of a suchness that is the Qentinuum of limitless energy and unformed, infinite conscious intelligence from which this and all other Multiverses emerge and evolve~ If one could somehow step into Qen outside of the Multiverse and look back upon it, one would see trillions of galaxies ever-expanding into apparent nothingness, but nevertheless finite~ For this Thojourn, visualize that they would see this *nothingness* as an endless Intellergetic Qentinuum surrounding this Multiverse and at the same time, permeating every aspect of it~ Envision a painting in process that will never be complete, unfolding upon an infinite canvas~

Qen responds to *Intention*~

Your reality is now at this moment being created by your intention, awakening Qen and Intellergy Forming the next *Now* of reality for you~ The overwhelming bulk of this intention is unconscious, reactive, and predictable~

You possess the free will and choice to become mindful and proactive in this Intellergy Forming process~

Qen is responding at this very moment~

What would you like your next manifestation to be?

I'm feeling good 'cause I want to~
— The Tortugans

Choose Qen as your favorite toy~

Lotas by Kelci Auliya

XV

The Indigos

In the Terra Signature at this time, many physicists, philosophers, and mystics are recognizing a transition that is being called The Great Shift~ What are being known as the *Indigo, Millennium,* or *Light* children are beginning to experience conversations and transmit personal contact as exchanges of energy~ They are beginning to leave behind the idea of a vulnerable and finite body, and they are Intellergy Forming their new reality as Energy Signatures~

The millennial Indigos are finding themselves in a unique position, being raised and educated by mind/brains that are not as evolved as their own~ These Indigos are demonstrating *enhanced* pre-frontal cortex activity levels, which avail them of similar advantages their ancestors held over the Neanderthal, such as accelerated thought processes, creativity, and adaptation to new technologies~ Many of these youth are being diagnosed with Attention Deficit, Hyperactivity or Neural Development Disorders~ Their parents and teachers become frustrated with them because of their heightened energy levels, reluctance to adapt to the *old ways,* and their rebellious reactions when they are forced to conform~ These millennials do not readily respond to traditional education or discipline~ Unfortunately, many are being drugged, punished, or ostracized as *weird*~ As a result, too many have resorted to substance abuse, violence, or depression with even suicidal tendencies~ Although there is an overwhelming amount of resource material from global sources offering ways for communication, education, and enabling this new mind/brain, the vast majority of teachers and families have not even become aware this *shift* is occurring~

Imagine in seventy-five years, when the children of these Indigos have reached adulthood, with life expectancies of 150 years or more and access to the information and resources available at that time~ By this stage, carbon-oxygen based Intellergy will have seamlessly integrated with quantum computer carbon-silicon Intellergy to create the next manifestation of intelligent beings~ It is fairly simple to imagine that the reality of those days will little resemble those of our present~

Blindsight absolutely compels that Energy Signatures who are able to recognize this shift begin to celebrate and facilitate this amazing evolutionary period~ This is simply the most important time of change in the history of the earth~ Even though most likely none of us who are possessed of pre-Indigo intelligence can even begin to comprehend this new reality, it is most certainly exciting and pleasurable to watch~

~~~*~~~

For some better understanding for those of us in this Great Shift transition, here is some relevant history~ About 550,000 Earth years ago, Neanderthals walked the planet~ They possessed language, art, music, tools, weapons, religion, and history~ They were never self-destructive and in fact lived within a reality structure that embraced, inasmuch as they could understand, all of the forces in their Terra Signature~

Then, without any warning, Homo Sapiens appeared on the field, with a much expanded pre-frontal cortex enabling much more cognitive behavior, creativity, and long-term planning, much like our millennial Indigos in relation to us~ There is no compelling evidence to suggest some kind of ultimate Armageddon of the species that could have led to the Neanderthal's demise~ Rather, it is more evident that these new sentient bipeds were able to create better tools and weapons for hunting~ They were able to fashion durable and portable shelters, enabling them to expand their hunting and gathering ranges~ They were able to develop agriculture for food, medicines, and stability~ All in all, they forged a completely new reality~ The Neanderthal was not destroyed, they merely became irrelevant~

An interesting thought experiment might be to imagine one of these new Sapiens being raised in a Neanderthal cave family~ This human would literally neither understand nor ultimately respect the old paradigms such as their history, religion, educational systems, social structure, or philosophies~ They would fundamentally rebel against ideas and practices they found absurd in the context of their new vision of reality~ Imagine how the Neanderthal would react to this rebellion~

(For a fascinating and meticulously researched read portraying this scenario in the richest of detail, the reader is encouraged to find Jean Auel's Clan of the Cave Bear *along with the sequels in her* Earth Children *series.)*

~~~*~~~

Energy Signatures afflicted with the Human Disease are the Neanderthals in the age of The Great Shift~ They are either ignorantly oblivious to the lightning evolution that is occurring around them or they are defensive and obstructive~ Sadly for them, they will find it completely unstoppable~ It is very likely the Neanderthals never saw it coming either~

One Mind is driven by creation in the pursuit of ever-increasing awareness and intelligence to better observe itself~ It is absolutely unmoved by the survival of a species if it is not moving toward this goal~

Carl Sagan, the prolific astrophysicist, astrobiologist, author, and science communicator, reflected that almost 99 percent of all species that ever existed on Earth are extinct~ He wrote, "Extinction is the rule~ Survival is the exception." It is fair to say no species ever plans to disappear~

Already there are Indigos at the age of sixteen and seventeen years conducting classes, posting blogs, and producing web casts on Quantum Mechanics, engaging in Intellergetic conversations, and living as energy beings~ Those who insist on remaining in a physical body perspective will simply find themselves obsolete, too slow to react, and unable to understand or relate to the new paradigm~

The cure for the Human Disease will be found for those who find themselves able to first realize and embrace the fact that they have the condition and then engage conscious, proactive Thojourns, such as creative visualization or meditation to begin treatment~ The immediate commencement of vigorous, intentive self-observation is essential to start revealing all derivative, formulaic behavior and language, as these are the centers of attractive density that are constantly cycling in and upon themselves, building more and more blockages and accumulations in the energy sphere~ Several scenarios and suggestions are offered in this narrative; however, creating a personally designed path based on individual experiential densities found within one's own Energy Signature will most often lead to more accelerated, profound and long-lasting re-sults~ _____ loves to play and create~ By embracing and celebrating these characteristics, there is great enjoyment and adventure to be had~

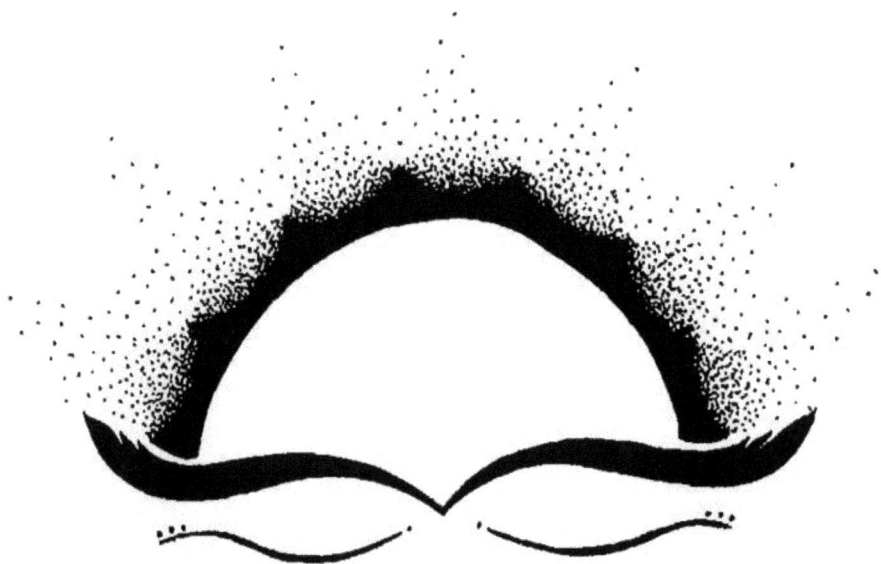

XVI

Learning How to Learn

*Formal education is
a walk through the zoo~
Informal learning is a walk through the Savannah~*
— Stephen W. Hart

*Tell me and I forget~
Teach me and I remember~
Involve me and I learn~*
— Benjamin Franklin

*It's what we think we know
that keeps us from learning~*
— Claude Bernard

No one can teach anyone anything~
Even the words *teacher* and *educator* are misnomers~
One can only learn~

The very definition of a student is that of a learner~ Perhaps one should call teachers *Learning Facilitators, Student Enablers,* or *Pupil Initiators~*

Legislatures, Teacher Conferences and, with few exceptions, all their members ascribe to a fundamental approach assuming students can be taught~ This is not a categorical criticism of what we consider arguably the most important endeavor in the Terra Signature~ We would hope it will be accepted more as a challenge, with the understanding that most of those in the field are constantly searching out and experimenting with ways to inspire and focus their classrooms~ What is offered through an Intellergetic awareness is a different, albeit simpler, perspective~

The greatest and most exciting challenge of the millennium is helping the new Indigos maximize the potential of this innovative Intellergy~ It is becoming dismally clear that traditional Sapiens educational methods are simply too slow, insubstantial, and linear for the Indigo Energy Signatures; and yes, it is boring for them~ The cry should be for what we would call *Evolutionary Learning*~ This movement should be motivated observationally by allowing the Indigos to demonstrate to us how it is they learn~ The pre-millennial intelligence operates from different frames of Qen4 reference, and as a result, the structures, materials, and protocols are driving current educational cultures to record dropout rates~

The title of our chapter is Learning How to Learn, and from a Blindsight perspective, the process should be fairly simple~ Fundamentally, only two things are required of a sentient being for learning:

Focus on the subject at hand for extended periods~
Remember what was observed or experienced~
Focus and *Remember*~
Fairly simple don't you think?

One only has to walk into a millennial toy or baby store to quickly see how far we are removed from simplicity in terms of creating environments conducive to these modest learning skills~ The modern marketing machine would have the parent of a newborn or a child of school age believe that their bedrooms and classrooms should consist of an elaborate complexity of imagery, objects, and sounds~ It becomes the goal to *stimulate* the child as much as possible in hopes their burgeoning intelligences will respond with absorbing all the intricacy, density, and cacophony of the millennium in ways that will help them excel in a world constructed in this way~

One might imagine a typical millennial infant's room~ Mounted over the crib would be the obligatory mobile of colorful and complex characters conveying all sorts of energy and attitude, along with the accompanying music theme of the day~ Naturally, walls and ceilings of the room have been either covered with murals or hung with art and posters~ In the closet and on the dressers, tables, and floor lie hordes of toys, media, and games, all awaiting choice, intention and participatory activity~ The parents engage with accelerated energy and voices of higher-than-natural pitch in order to stimulate reaction~

This brand new, fragile, amazingly intelligent, intuitive Intellergetic Being is open to absorb any and all input, store it, process it, and prepare to transmit, for this is one of the newest apertures for _____'s observation of itself~ The reality of this process is that immediate attachment to complexity, confusion, and density is being instilled in the Energy Signature~ With focus and attention constantly being diverted and distracted while memory is rarely encouraged, the young Intellergy trains itself to engage its perceptions more on a comprehensive, multitasking scale rather than one of quality and single-mindedness~ Duration in this broad approach to awareness is completely compromised, as the young Intellergy is constantly and repeatedly distracted and then attracted to the next impression or activity and the next~

Unfortunately, these Indigos are met, at least in most of the existing schools, with classrooms that reflect exactly this same environment of complexity and relentless activity, with almost no attention to focus or memory training~

Blindsight would propose that if one learns at a very early age how to focus on one object or activity for extended periods of time while practicing how to remember what it is they have experienced, this Intellergetic Being should be able to apply these skills to learning virtually anything: math, science, languages, art, technology, literature or any subject~

Imagine an Intellergetic infant's room where simplicity reigns, with subtle artwork more abstract in nature and softer in color tone~ The music in the space is instrumental with no lyrical content in a slow, relaxing tempo to facilitate alpha states in the mind that have been proven to increase learning speed and retention~ The baby gets only one toy at a time to play with and focus upon~ Please understand this is not punishment or withholding playthings from an Energy Signature designed to perform at this fundamental, simplistic stage of its evolution~ This can be any type of object, not necessarily one from the mega toy superstore~ This expanding Energy Signature is, by its very nature, curious and exploratory and will immediately begin to focus its entire attention and intention on its new *toy*~ We have personally watched this experimentation go on and on with infants literally for hours in calm, intentive play~

At the next play opportunity, a completely different play object may be introduced in a different size, weight and texture~ This has a two-fold purpose~ First, it minimizes attachment densities and second, it expands Intellergy by broadening experiential awareness~ For example, one session might engage the child with a stuffed animal, the next with a shiny piece of quartz, the next with a ball, and the next with a smooth block of wood~

By carefully observing the child if they begin to tire of the toy, it is helpful at this point to engage them in small memory exercises~ In such a case, one might get three small pillows~

Have the child focus on the stuffed animal if that is the toy of the moment, then show them that you have hidden the toy behind one of the pillows~ Now, begin to move the toy behind different pillows, letting the child focus on the motion, then remove yourself and let the child find the animal~ You will find ways to increase complexity with this type of game as the young Indigo progresses, and you will be amazed at the acceleration in focus, memory, intellect, creativity, and self-confidence that emerges~

As the child grows, you may create your own focus and memory games engaging opportunities at meals, baths, bedtime, traveling, and many other places~ In fact, once you and your child have more experience with memory and focus games, you will find they can be played anywhere and anytime~ There is also a wealth of children's memory games available to expand your repertoire~

Imagine a child of four entering preschool with total confidence in their abilities to focus on and memorize everything they encounter~

What is the potential of an Intellergetic Indigo child with this skill set?

In concert with the focus and memory games, it is important to engage in building hand/eye coordination immediately~ This is so simple and fun for both parent and child~ As the *focus/memory session* begins to wind down, have the child sit facing you and hold the object or toy they have been focusing on where they can clearly see it~ Use movement to animate the toy so the child is concentrating on the toy and not you~

It is important to resist extended eye and vocal contact to gently lead the child toward focus on the activity, not the adult~

Gently and slowly move the toy into their hands and let them grab it~ With kind and relaxed enthusiasm, tell the child how well they did and suggest trying it again~ With repetition, this game will quickly progress in speed and skill, reaching the point where the infant is actually catching the object with confidence~ Encourage the child to toss the object back to you~ By attempting to get them to use alternating hands and feet in this simple game of throw and catch, you will not only stimulate different areas of Intellergy, but it will also foster the beginnings of ambidextrousness, opening neural pathways for more fluid and improvisational intelligence~

Please do not underestimate the ability of an Indigo child with such a simple introduction in learning to manage the accelerated complexity of the millennial Terra Signature~ On the contrary, such Intellergy will far outstrip the attention and memory-deprived Energy Signatures in their learning spheres~ These Indigos' expanded ability to immediately grasp concepts, store them in their fluid, infinitely expansive memories, and process them effortlessly at extraordinary speeds for transmission is exactly what their evolved millennial brains are designed to do~

Blindsight hopes that the millennial parent will appreciate the aspect of play that is infused into every aspect of the early Indigo's learning experience~ Remember always, One Mind is very loving and playful, celebrating and rewarding all intention rooted in this spirit~

The relaxed and uncluttered environment creates the space for unlimited XPotential creativity that a loud, fast, and complex atmosphere could never begin to provide~

In the writing of this book, we have come to notice the interactions between parents and children more often than before, and it has been remarkable to note parent response to the apparently unending flow of questions that begin to emerge in a child of two to three years of age~

Why is lightning white, Daddy?
How do they make chicken nuggets, Grandma?
Why do I have hair, Mommy?
What does dead mean?

All these types of questions are asked with total and open innocence and the expectation of a reasonable answer~ In virtually all our observations of these amazing and wonderful moments, the loving parent, grandparent, friend, or adult du jour makes the most valiant and sincere effort to answer the question in such a way that the child can understand~ On the more difficult questions, such as the last one we offered and which we actually overheard in a store, the adult is most often visibly uncomfortable, hesitating and stuttering while they struggle to craft just the right response that will at once answer the question, satisfy the child, and not confuse or frighten them~ Adults seem to want most often to see themselves as the person of knowledge, authority, and responsibility~ Very little, if any personal power is awarded the very young~

How different the scenario if the response had been phrased in the form of a question~

I have some ideas, but why do you think lightning is white?

That's an awesome question; how do YOU think they make chicken nuggets?

We all have hair; what do you think it's for?

What does "dead" mean to you?

Now here is the hardest part: **WAIT FOR IT~**

Allow the child to take in this challenge, store it, process it, and transmit when they are ready~ There is absolutely no hurry here, and often the question alone will divert the Intellergy;

however, most of the time a child will come up with the most creative and imaginative scenarios for their answers~ Please allow yourself, the adult, the opportunity to have some fun and support these new realities and Intellergies without question or criticism~ The very basis of all invention, art, and entrepreneurism is unencumbered, fearless, innovative thought~ As mentioned in the early chapter on Quantafiers, all things in human existence begin with thought~

The value of these responses and those like them stems from the freedom and power afforded the child as they learn to immerse and understand their Intellergy in their true nature of infinite potential in consciousness and energy~ The more we respond to children with questions and allow them the time to formulate responses that are creative and right for them in that space, the more confident they become in their abilities to comprehend and approach the greater questions and challenges as their manifestation evolves~

It is also important to celebrate any response or answer that the child offers, not necessarily with exuberance, but with controlled exaggeration and unconditional loving energy~ Most often in the very young, the replies may take on outrageous connectivity to other concepts that have absolutely no relation to the original question~ This is simply so very beautiful~ Allow for the stream of consciousness to flow and enjoy the ride~

It is estimated that a child hears the word *No* in some form or fashion thousands of times before age four; *Stop that, Don't do that*~ This has traditionally been the method of protecting the youngster who might fearlessly stick their finger in the electric socket or blindly walk into the street~ Sadly, *No* is also a way to control a child while the adult attempts to manage an already complicated agenda and schedule~ Unfortunately, *No* often

steps in the way of surprising and amazing opportunities to facilitate learning, awareness, observational and physical skills~

~~~*~~~

*Our oldest daughter was about eighteen months old
when one evening we found ourselves having dinner in
a nice restaurant~ She had proven to be one of the early
Indigos with astonishing sensitivities, working presently
every day in the creation of a Pod of Light Bearers,
networking with others around the world to bring about
change in the Terra Signature~ As a result, she was, to say
the least, quite a handful~*

*She had chosen this particular evening to act on
this extraordinary acumen and energy, with constant
attempts at exploration in the establishment~ Her protests
continued to rise in volume each time her mother or myself
attempted to distract her from venturing out~*

*It became apparent that many of the other guests
in the room were not only beginning to notice, but also
becoming irritated at the disturbance~ We had often
witnessed parents in similar situations speaking harshly
to the child, slapping a hand, threatening ("Wait until we
get you home!") or removing the child from the room~
Instead, we picked her up, held her close and told her how
wonderful and amazing she was~ Then we simply asked
her to look around the room and observe that a lot of people
were looking at her~ We asked her how she thought they
felt about her being loud and trying to run around their
tables~ The look on her face spoke volumes as she instantly
became present and mindful of not only her own expanded*

space, but also intimately aware of her entanglement with others in the room and their reactions to her behavior~ She immediately went back to her seat, having made the conscious, mindful choice to end her performance without any adult making the decision for her~ She happily and quietly finished the meal with us and had the best time~

From that time forward, we were able to bring her anywhere without incident~ We actually used to comment that we could take her into a china shop and not be concerned~

~~~*~~~

The lesson offered here is simply that we can learn to trust our children no matter how young~ If they are freely given the opportunities to interact with their world, having been first awakened to their NOW Intellergy, they will almost always make creative and positive choices~ In our scenario, negative language or discipline would have almost assuredly created densities attracting more of the same kinds of behavior~

> *Help people reach their full potential~*
> *Catch them doing something right~*
> — Kenneth H. Blanchard,
>
> *Whale Done*

In Blanchard's wonderful book referenced above, he explores the relationship between the Orcas (Killer Whales) of SeaWorld with their trainers and compares them with personal human and management scenarios~ The trainers were very quick to point out that it is all but impossible to get a twelve thousand-pound predator to do anything if you punish or abuse them~ Only by finding the whales doing something right and then rewarding that behavior does progress toward positive performance emerge~ They also noted that negative or rebellious actions rapidly decreased as the whales learned that pleasing the trainer brought more and greater rewards~

~~~*~~~

Several years ago, we inherited Lola, a hundred-pound dog of Great Pyrenees and Shar-pei mixture~ Both these breeds are known for their fierce independence and stubbornness~ This highly intelligent and overly friendly giant had been spending most of her days locked in her owner's apartment while they were off to work~ In her two years of life, she had chewed up or actually eaten most of the owner's shoes, clothing, and furniture~ In her morning and evening walks, she had dislocated the owner's shoulders on two occasions, not out of any maliciousness, but simply out of enthusiastic opportunities to use her magnificent energy and strength in pursuit of another dog or a cat~ Up to this point, Lola had been disciplined with swats of newspaper, leashes, and the like and was apparently impervious to pain of any kind~

We offered to give her a home on our modest three-acre mini-ranch where she would have at least room to roam~ It is notable that from day one, Lola never chewed up anything except the bones we would give her from time to time~ Having the opportunity to expend her enormous wealth of energy on our beautiful property was life-altering for her~

During this period, we were exercising ourselves by walking from one to two miles each day, so it became a natural progression to put Lola on the leash and have her join in~ We had had some experience in training dogs and came from a school where the trainer takes on the role of the Alpha male in the pack~ This method speaks to the pack mentality in the animal that is ingrained naturally and instinctively~ In the wild, an Alpha male is very aggressive in nature to protect and evolve his pack, so the training has these types of practices~ It engages in quick, sharp **checks** with the leash to return focus to the animal and direct behavior~

We would like to remind the reader here that Lola was virtually impervious to pain~

On our first walk, she saw another dog at a fence across the road~ Before we even had time to react, she had bolted, dragging us over the road irrespective of any oncoming vehicles, tumbling us down into a four-foot ditch on the other side, up the embankment and into a barbed-wire fence~ Bruised, twisted, and bleeding, we lost all presence and began swatting her nose with the leash until she finally stopped her barking and fence fighting~ As we sat in the grass regaining our breath and composure, Lola lay calmly, panting and tongue-lolling in total bliss~ Suddenly, we remembered the treats we had put in our pockets before we left the house~

We checked her gently into a **sit** position verbally giving the command and gave her a treat; insert tail-wag here~ We began walking and after a few steps, we gave the command first and after about two seconds checked her into the sit position and gave her a treat~ The third time we stopped and gave the command, she sat on her *own! Eureka! Epiphany! Lola is a treat hog, and a highly intelligent one at that~*

Within two months, Lola was responding instantly and happily to hand signals for all basic commands~ Today, when walking whether on leash or freely, she will always remain calmly at our left side and the leash never has any tension~ Treats are always handy~

All of this is the result of finding her doing things right and simply rewarding that~ What would be the potential of a parent/child relationship founded on this principle?

~~~*~~~

How many times have we witnessed parents disciplining their children in public, up to and including striking the child?

What is the child learning from this experience?

The first lesson dynamically ingrained is that violence is the simplest, fastest solution to resolve conflict~

Secondly, the child learns that respect for your parent or guardian is fear-based and not love-based~

Tough Love may get results; however, the densities and attractive harmonics that are created most certainly lead to unforeseen cycles of future negative behavior in relationships following the Energy Signature sometimes for their entire manifestation~ We have often remarked that if any adult ever feels they have to strike a child, they are fundamentally admitting they are not intelligent and present enough to create another solution~

Learn how to learn~ It is so simple: focus and remember~

Immersion into _____ is so simple~

Immediate access to infinite Intellergy is so simple~

This is one's true nature~

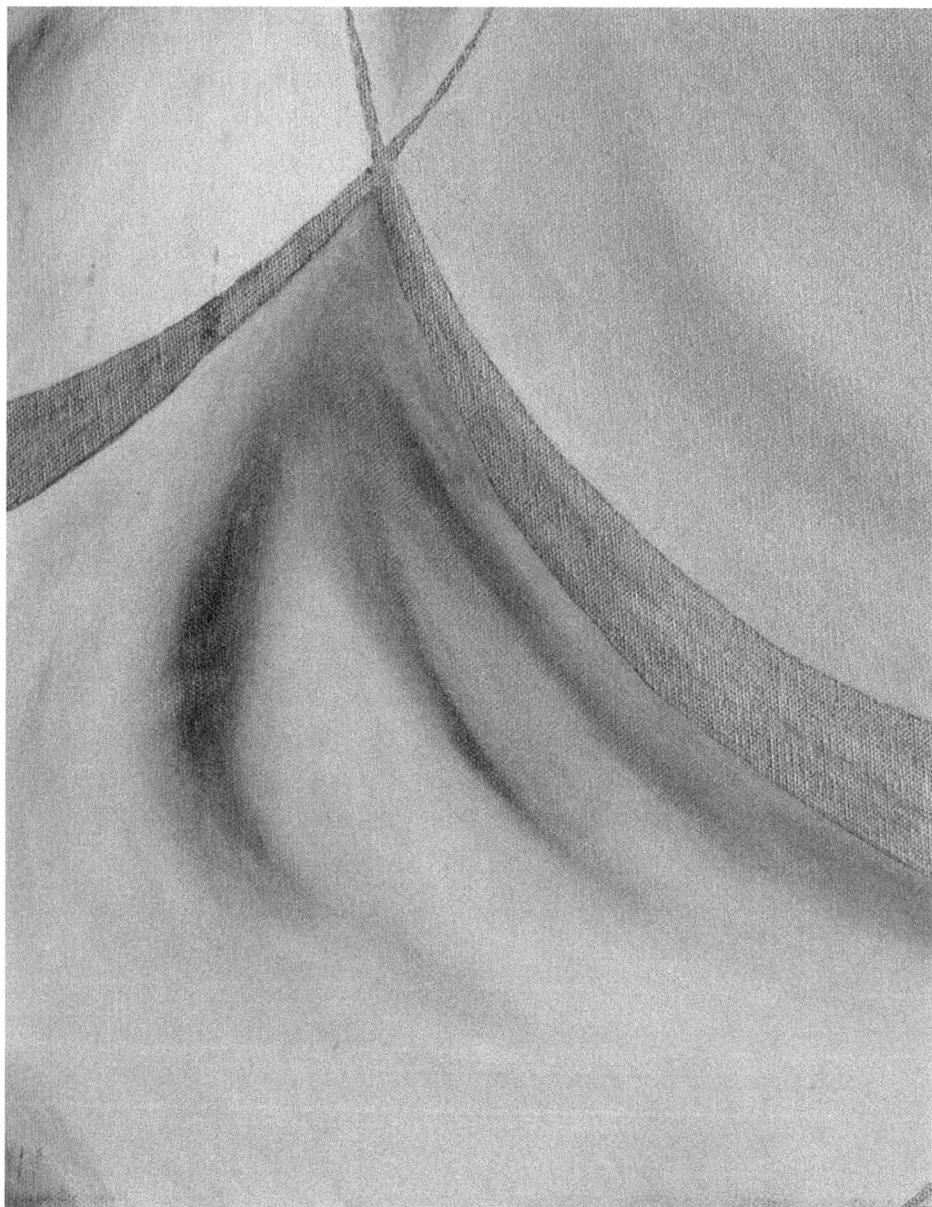

Firinn by Erin Nicole

XVII

Qensional Bridge

All the words and page simulations herein are Qen4 constructs~ Language possesses height, width, and depth~ Words expressed, whether spoken or in book form and other media, are capable of movement through time~

At the same time, they have the power to convey energy and self-aware information, bridging with Qen5 and higher sources of intelligence and perspective~

It is the goal of Blindsight to contribute to the building of this bridge through this simulation and some of the new language herein~

Here is an oversimplified Blindsight perspective on Qen harmonic strata to provide a basis for this chapter:

Qen1 is a Singularity~ It is a quantum position of infinite potential~ Mass, Time, and Space do not exist in a Singularity~

It is from a singularity that the Big Bang emerged~ It has also been attributed to the center of Black Holes; however, we will not be discussing that particular reference here~

Qen1 has no height, no width, no depth, and no movement through time, yet it has existence~ It also represents infinite XPotential of energy and consciousness~

Qen2 could be imagined as an essentially flat Multiverse possessing length and width, but no height or time motion~ A Qen2 being could never conceive of a Qen3 existence involving height as separation from the plane or movement through time~

Qen3 is height, width, and depth, like that of a cube with no movement in time~

Qen4 is time; Qen3 movement through time~ Although we are actually Qen4 beings, we are no more able to see from a time perspective than a Qen2 being could see as a Qen3~ If one could *see* as a Qen4, one would be able to experience every event and object with its entire evolution in time~ Oversimplified, one could look at a table and experience its construction, the tree that the wood came from, as well as the seed that grew the tree, the previous tree that dropped the seed, and on and on~

Qen4 is a realm of duality~ The movement through time creates the illusion of separateness, motion, and division with its perceptions of you versus me, here versus there, good versus evil and so on~ Duality complicated by the attempts of language to control, define, and qualify its manifestations remains a fundamental cause of all conflict, tension, and misunderstanding in this Qen stratum, resulting in the Human Disease~ Energy Signatures find it extremely difficult to remain entangled with One Mind in this dimension because of accelerated, excessive, and dualistic turbulence in the probability cloud~

Qen5 has often been referred to as the Dimension of Light, as it exists beyond the constraints of a Space-Time Continuum~ This is a realm of so-called Enlightenment, Samadhi, Nirvana, Satori, Heaven, and the like~ Many shamans, mystics, healers, prophets, and other *sensitives* are able to interact on a Qen4 level based on the energetic intelligence of Qen5~ Remarkable insight and manifestations can emerge from such an interaction without Qen4 limitation, and it offers a glimpse into the possibilities and unlimited potential of an Energy Signature as a Signome~

At the level of Qen5, it is fundamentally impossible for the Qen4 concept of evil to exist~ In a realm where there can be no duality, constructs such as fear, misunderstanding, possession, conflict, pain, or tension; only beings that are free of these densities are welcomed even for short periods and allowed to remain while further challenges wait for transition to ever more expanded reaches of space and immersion into _____~

During the transitional phases from Qen4 to Qen5 realities, there are countless events when Intellergies find themselves in the Nirvana of Qen5 space, only to be attracted back into Qen4 instability by unresolved dualities and densities within the Energy Signature~ However, each glimpse into the possibilities that await for a purely Intellergetic being freed from Qen4 turmoil provides inspiration and renewed encouragement to pursue pathways through to fully expanded space~ An Energy Signature is fundamentally altered harmonically by such an experience, with density centers relieved of many accumulations and dualistic tensions~

Blindsight is a Qensional Bridge constructed as a humble pioneering effort to open passages to Qen5 perspectives~ Just like early pioneers and explorers struck out into uncharted lands, seas, and worlds, Blindsight strives to create Thojourns to reach new and wondrous realms with the possibilities of connecting with not only similar resonant voyagers, but also those with far greater Intellergetic harmonics~

Blindsight recognizes that earthly human Energy Signatures are merely the first (or perhaps current) intelligent beings manifested in this tiny *corner* of the Multiverse capable of self-awareness~ This would imply that these Intellergies are able to observe themselves and fashion tools of language, mathematics, and science to refine and expand their observations~

Metaphysics, religion, the occult, mysticism, and the like have forever alluded to the existence of angels, ghosts, spirits, oversouls, guides, fairies, and the like~ For most, these are seen as superstition, hallucination, or delusion~ Those who actually experience these connections find them as real as a conversation with a friend~

There are circumstances in which transitional Intellergies also take on the flow and momentum of thoughts, actions, and events at the time of transition~ This is more common in sudden transition events, such as accidents or abrupt failure of the physical body~ Blindsight makes allowances that certain Energy Signatures under these extreme types of circumstances can find certain aspects of their manifestation *stuck* Intellergetically when the flow of their transition is interrupted in too immediate a fashion~ This can be imagined much like a whirlpool occurring in an outcropping such as a sandbar in the middle of a flowing river~ These *density beings* in no way represent the entirety of their Intellergy, just as the whirlpool in no way represents the entire stream, but nevertheless they retain certain identifying characteristics that have become somehow partially severed harmonically from the flowing source in a repetitive cycle ~

In the schools of the occult and the paranormal, many have come to call these density beings *ghosts* or *sprits,* and because of simple misunderstanding of their true nature, intention that is fear-based has been leveled at them, attaching even more density to their signatures, prolonging their manifestations and binding them to Qen4 frequencies~ Inevitably, just like that wave that returns to the ocean, these beings will succumb to this entropic and temporary aspect and become reabsorbed into the Multiverse~ No Intellergy is ever lost~ There are certain Light Bearers with frequency *tuning* abilities who are able to resonate with trapped Intellergies such as these~ With harmonic intention, they are able to resonate energy frequencies, breaking them loose and releasing them from their cycles~

The Entangled Density Principle of Blindsight compels that there is an almost limitless number of resonant or harmonic Intellergies *linked* with every Energy Signature~ These could be looked at as the Oversouls alluded to by Seth in the Jane Roberts dialogues, J. Z. Knight's Ramtha or perhaps the *entities* or *guides* of John Lilly~ We might refer to them as Meta-signatures~

Each Energy Signature is harmonically in resonance with much higher Intellergies~ Meta-signatures rarely communicate in Qen4 language or imagery, but are rather perceived as impression, epiphany, or inspiration~ In Blindsight, it becomes an imperative that these Intellergies are welcomed and employed as vehicles for transition into higher, more expanded space~ These resonant Intellergies lie openly in wait for Qen4 beings who allow themselves to become channels for these evolutionary steps~

From an earthly perspective, our solar system is about four billion years old in a 13.7 billion-year-old universe~ Sapiens intelligence appears at this writing to be about two hundred thousand years old~ Look how far it has come in that relatively miniscule length of time~ Now imagine an Intellergetic Being that is five hundred thousand years old~ What about a million years? A billion? Thirteen billion?

If such beings would ever have any need to communicate at our level, are we really so naive to believe they would show up in a disc-shaped object with anal probes and ray guns, speaking an earthly language? It is not beyond the realms of imagination that our Energy Signatures will be capable of this type of exploration within a relatively short period of time~ It would seem the anal probes might be a little excessive~

Certainly from a Blindsight entanglement point of view, all Intellergetic beings are entangled in the same multiversal intelligence, no matter our age in space-time; ultimately, the idea of other beings or intelligences is illusory~ There is only one Intellergy and that is One Mind~

If you have not begun to feel the Feynman *crazy* by this time, you may begin~

Nevertheless, it is our hope you are ready to cross the bridge~

> *Life is a bridge~*
> *Cross over it,*
> *but build no house on it~*
> — Indian Proverb

NOTES

Sinnsear Craobh by Kelci Auliya

XVIII

I Think

On the night of St. Martin's Day, 1619, in Neuburg an der Donau, Germany, Rene Descartes, the French philosopher, mathematician, and scientist had found himself in a room with an *oven* to gain refuge from the cold~ Over the course of the long night, he had a series of *visions* that later went on to evolve into a core philosophy that not only defined Descartes' life, but also the essentials of his philosophical works~

In his *Discourse on the Method* published in 1637, he wrote, "I think, therefore I am." He defended this logically with principles such as the fact that even "if one is skeptical of existence, that is in and of itself proof that he does exist."

Many would offer at this reading that this would be a ridiculous waste of time~ Of course I exist! Why would one even question such a thing?

So we did~

One evening, deep in a meditation on _____, we found our self in a real and profound realization that I AM all things~ This was an *Impersonal I* experience, not a personal *I*~

There was no duality~

No you and me~

No here and there~

No this and that~

No wrong or right~

No heaven or hell~

No zeroes or ones~

We became what Quantum Mechanics calls *entangled,* a state where no particle (or entity) can be described independently~ Any quantum state can only be described as the *system as a whole~* In Quantum Mechanics, there is an integral piece of every calculation, experiment, and theory that must be included and cannot be ignored: That is the role of the *Observer,* even though this Observer has never been observed~

> *The observer is the observed~*
> — Jiddu Krishnamurti

Ah, Mr. Feynman~

Over the last century, countless experiments and theories have been tried to identify, define, measure, or observe the Observer to no avail~ Every attempt ends in the inevitable collapse of the wave function performing a selective outcome~ Each time, the event dissolves into duality~ It is a *particle* or it is a *wave;* the cat is alive or it is dead~

Albert Einstein was even heard to ponder whether the moon was actually there when we are not looking at it~

In our meditation, the *Observer* is _____~ _____ is the source, the creator, the realized, sempiternal, multiversal self.

> *I AM the nature of pure consciousness~ I AM the same to beings, one alone; I AM the highest Brahman which like the sky, is all pervading, imperishable, auspicious, uninterrupted, undivided, and devoid of action~ I do not belong to anything since I AM free from attachment~ I AM the highest Brahman . . . ever-shining, unborn, one alone, imperishable, stainless, all-pervading, and nondual~ That I AM, and I AM forever released~*
> — Shankara, *The Upadesasahasri*

Shankara is speaking in the *Impersonal I* that is _____ exclusive of any form, object or action~

From this meditative *Impersonal I AM* perspective, our own oneness suddenly and completely unexpectedly became a state of deep and abiding terror~

We became fundamentally aware that everything we had always observed, experienced, and known was only illusory~ We were completely alone in the Multiverse, with no physical existence or so-called life~ This point of view demanded that such relationships with our wife, children, passions, work, everything we thought we knew, were merely manifestations we had created: brief, fluid, fragile, and ultimately unreal in any kind of reality other than one of our own creation~

We were nothing~

We were completely and totally alone~

Then the most wonderful thing happened~

There was the present and clear knowing that *I AM* (impersonal), at the same time everything *and* nothing~ This was Franklin Merrell-Wolff's *Consciousness without an Object*~ It was the Nirvana, the Samadhi, or the Satori of the ancient mystics~

There is no fear~

There is no death~

There is no misunderstanding~

There is no time~

There is no motion~

I AM unconditional love~

I AM infinite XPotential of undefined energy and consciousness~

It is fair to ask at this point, why keep writing?

Merrell-Wolff wrote literally hundreds of pages on *Consciousness without an Object* after he wrote in Chapter 2 that language "ceased to have any validity."

Go figure~

If we cannot even be assured that there is even anyone out there to read this other than ourselves as the Observer, then perhaps we are merely creating this Multiverse in which we are merely a gnarly, old Energy Signature rumbling around in a One Mind, feeding our own illusion with unending manifestations for our own amusement~

So be it~

If _____ is the ultimate loneliness, then perhaps _____ is forever creating Multiverses of infinite and random natures to entertain or interact with itself~

From a quantum perspective, a Multiverse responds to only one thing: Intention~ Whether a pure act of _____ immersion or a reactive, harmonic resonance *(See the chapters on The Multiverse Guitar and The Qen Trinity)*, infinite universes, and dimensions within them evolve and devolve, expand and contract at the whim of intention~

Nothing is lost~
Nothing is gained~
Breath of Brahman~
Ultimately, _____~

NOTES

Gaol by Erin Nicole

XIX

Intellergetic Sex

This is probably the chapter you have been waiting for; it's about time someone took Tantra to the quantum level!

We should apologize that Blindsight would offer a slightly different perspective~

It seems worthy at this stage in the evolution of this particular Terra Signature to note the significant acceleration and density of human male/female Intellergy in the millennium~ With populations growing apparently unchecked at exponential rates, this turbulence and density is the main impetus pushing us toward the inevitable Quantum Point ~

Over the millennia, this point has been foretold as Astrology's Dawning of the Age of Aquarius, the Hopi's Return of Kokopelli, the prophecies of Nostradamus, John of Patmos' Armageddon, the Second Coming of Christ, or the Great Shift of the quantum physicists and millennium philosophers~

~~~*~~~

One curious sidebar with respect to the above predictions originates in the pseudoscience of astrology~ Long considered a fragmented science, most of its fundamental and connective tenets were lost over the last four thousand years or so~ Nevertheless, in the millennia

*BCE, predictions were made that a ninth planet would be discovered at the beginning of the Aquarian Epoch that would bring Virgo into its full potential~ It is interesting that recently Pluto was demoted from full honors as a planet, leaving our solar system with only eight~ Pluto has been reclassified as a **dwarf planet**~*

As of this writing, Mike Brown and Konstatin Batygin of Caltech have produced evidence of such a ninth planet that would roughly be ten thousand times the size of Earth~ They have found a gravitational effect that is pulling smaller objects into clusters, indicating this planet lies very far away, as much as one hundred billion miles on the other side of the sun~ Its solar orbit, most probably in the darkest outside loop of its path, could be as much as fifteen thousand years, but it may be observable in the very near future~ This coincidental cosmological possibility is simply something else to watch for during the Shift and the approaching Quantum Point~

~~~*~~~

At the emergence of humanity, female energy was understood and often worshiped as the *source,* where male forces were absorbed, blended, and expanded for balance and harmony in the world~ Over time, male energy steadily pulled away from this balance, climbing to ever-higher peaks of domination and control, contributing to ever-increasing turbulence, tension, and density in the Terra Signature~

It is interesting also to note at this point in time the geometrical expansion of the Lesbian, Gay, Bisexual, and Transgender Queer energies on the planet, where the Intellergies of male and female begin to blur~ Male energy is being expressed through female form and vice versa~ Entirely new affectations of expression, dress, and culture are appearing~

Many heterosexual Energy Signatures are experiencing disturbance and resistance in their energy fields as a result of conflicting Intellergy that is perceived as a threat to their belief systems and pre-existing thought streams~ At the same time, many of these LGBTQ manifestations become so consumed by the forces of opposite sexualities within them, they search out and engage in transgender metamorphosis of their corporeal forms to facilitate the realization of the truer resonant nature of their Energy Signatures~ The social and political shock waves continue to escalate and reverberate through the harmonics of cultures in manifestations of demonstrations, celebrations, debates, confrontations, and congressional referendums~

From an Intellergetic perspective, this blending is an inevitable result of the enhanced density of human Intellergy Amalgams and hybrids such as these and many others will continue to manifest during the Great Shift in all other sectors such as economic, cultural, technological, biological, and virtually every other aspect of the Terra Signature, as Intellergy blends largely impacted by the incredible global mind of Internet access afforded Energy Signatures~ It has been said that a child with a cell phone herding cattle on the Serengeti holds in their hand more computing power than was available to President Richard Nixon during the Apollo 13 crisis in 1970~ This exponentially expanding global mind/brain connects all human Intellergy with loving, creative solutions, as well as darkly foreboding problems~

Each of these Energy Signatures, just like the rest of us, is infected with the Human Disease~ As of this writing, it is clear that there is an unavoidable Quantum Point approaching~ Humans, overwhelmed by this sheer contraction, are becoming more and more desensitized and insular in their understanding of self~ There would seem there are simply not enough actualized Intellergies rising up in such a way to inspire and challenge Energy Signatures to live energetically and entangled~

Thankfully, there are pods of Millennials, Indigos, and Light Bearers beginning to communicate, unite, and put forth strategies for the awakening fires of true energetic nature~ These Pods are Portals through which Qen5 Intellergy is flowing~ We urge the reader to search these groups out and learn ways to contribute to their spark~ One's intentions to find these Intellergies will create the proper attractions to the most harmonic resonances inherent in these pods and portals~

In Social Dynamics, there is a term taken from nuclear science called *critical mass*~ For Blindsight, it refers to a phenomenon where if enough people adopt an innovation or behavioral pattern, it can reach a point that is self-sustaining for the creation of further growth~ Blindsight would suggest that a critical mass of Light Bearer Intellergy could initiate a chain reaction, creating a fundamental shift to Qen5 aspects at the Quantum Point~

~~~*~~~

In the early 1990s, Washington, DC was experiencing drastically rising crime rates each year~

Between June 7 and July 30, 1993, during the hottest months of the year, there was a carefully controlled scientific demonstration study involving a group of Siddhas from the Transcendental Meditation tradition~ Beginning with about eight hundred practitioners, the numbers grew to more than four thousand over the period of the trial~ The study was led by John Hagelin, a renowned quantum physicist who predicted crime would be reduced at least 20 percent over the course of the trial~

The concept was to reduce stress and crime through mindful intention intervening from the field of consciousness~

An independent, twenty-seven-member Project Review Board was set up, made of independent scientists, psychologists, and leading citizens to create an indisputable research protocol able to withstand all challenges to fairness, bias, thoroughness, and results~

Prior to the study, local police actually ridiculed the project, noting that crime always rose during this hot time of year~ The Chief of Police was quoted as saying the only way to lower crime would be "twenty inches of snow."

The maximum impact was that there was a 23.3 percent drop in HRA Crime (homicides, rapes, and aggravated assaults) during the study~ It was interesting that the largest drop occurred in the final phases, when the largest number of meditators was in force~

Soon after the project, crime steadily began to rise once again~

The statistical probability for success in crime being impacted on the study was two in one billion~

After the study, David Edwards, PhD, Professor of Government at the University of Texas, wrote, "The potential impact of this research exceeds that of any other ongoing social or psychological research program~ It has survived a broader array of statistical tests than most research in the field of conflict resolution~ This work and the theory that informs it deserve the most serious consideration by academics and policy makers alike."

~~~*~~~

From this small experiment, can we begin to imagine the impact on the global Terra Signature with a confluence of higher frequency Intellergy and Intention reaching critical mass?

Unchecked, the Quantum Point will manifest in such a way that, within an extremely short amount of time, reality will appear as though *everything is happening at the same time*~ If one is honest about the acceleration of events just since the year 2000, it is not too difficult to imagine the potential~ This Quantum Point manifestation will be so strong, so overwhelming, that common human Intellergy will be unable to react~

If one gazes across the landscape of the Qen4 Terra Signature, one could imagine this Quantum Point occurring as environmental, cosmological, seismological, technological, biological, social, political, and personal all at once~ The outcome of such a series of events is completely unpredictable and would exist as a quantum probability cloud~ Like Schrodinger's Cat, the outcome is totally random until it is manifested and observed~ Universal Mind is at play; _____: The Observer~

Blindsight holds the belief system that with the self-actualized, fully entangled Intention of the Millennial Light Bearers, human Energy Signatures have the power to influence the outcome of the Quantum Point transition in positive and creative ways~ It is imperative that more Intellergies begin immediately to recognize their contamination from the Human Disease and begin to engage their passage across dimensional bridges to activate Qen5 intelligence~

There is simply no other viable alternative~ There are infinite options, but none would be nearly as much fun~

We have to understand that DNA energy and intelligence is blending within the literally billions of acts of sexual intercourse occurring daily in the Terra Signature~ Before the reader gets too stimulated here, please note that we speak of all animals and insects of air, land, and sea~ DNA has been found on meteorites that have fallen to Earth from deeper parts of the universe, so it would be extremely arrogant to assume that life on earth is all there is or even that it came first~ DNA appears to have an imperative of sorts that demands not just procreation, but also evolution, survival, mutation, and the ever-advancing search to observe itself through more evolved intelligence and energy~

~~~*~~~

Take a moment to embark on a short Thojourn, imagining the intelligence and energy exchanged in the act of human sexual intercourse resulting in conception~ Also for fun, remember as earlier discussed in the chapter on The Human Disease, nothing ever physically touches!

~~~*~~~

When sexuality and the DNA Imperative are viewed energetically from a Blindsight perspective, it becomes an even more compelling idea that we are approaching this true Quantum Point in the Terra Signature that was alluded to earlier and prophesied ages ago~ As to a positive or negative outcome of this Great Shift, it would be impossible to predict other than to understand that, with this acceleration of invigorated sexual energy transition, the Intellergies of the Indigo explosion, and the implications in speed and capacities of the global *brain*, one can be assured the Terra Signature will bear little resemblance to what has gone before~

Siorracd by Erin Nicole

XX

Neuroplasticity

Neuroplasticity, a term derived from the words *neuron* and *plastic,* refers to the ability of the mind/brain to adapt and respond to changing information for survival, maintenance, and creativity~

It was long believed that the brain could not generate new cells and neural pathways, but this has recently been proven to be false~ Research now shows that brains well up in age are still capable of cellular and synaptic growth~

Stimulation of this brain expansion can be increased by the learning of a new language or picking up a musical instrument, for example~ The more we practice and repeat a certain action stimulating brain activity, the faster the growth in skill and the better the retention~

The implications for Blindsight stem from a more energetic point of view~ Energy Signatures, by their very Intellergetic nature, are like a computer chip that takes in information, processes it, stores it, and then transmits it~ G. I. Gurdjieff often referred to humans as *radio transmitters~*

Energy Signatures develop individualities like *personalities~* Intellergetically these should be looked at as charged densities that continually attract other concentrations of like, similar, or harmonically related polarity~ In the densest of forms, these signatures actually become extremely attached to concepts such as their names, personalities, bodies, and experiences to the point these become belief systems that actually define them~ Some religious belief systems teach of realms or heavens where transitioning Intellergy actually retains these illusory *identities* living on as angelic beings, encountering past relationships and other mystical realities~

Taking the example of learning a language, each time the student encounters a new word or syntax, the energy is absorbed into the ever-increasing density bank of Intellergy~ The challenge for earthly Energy Signatures in humans has evolved to the point where the densities have become almost *crystallized*~ These energy *crystals* should not be looked at as actual crystals, of course, but rather very concentrated belief-energy systems fixed as preconceptions and expectations~ Once Energy Signatures take on this type of density, they can reach certain quantum points of their own that are resistant to Intellergy that is new or more fluid, and they tend to reject the new energy like a wrong note being played in a chord on a piano~ It might be experienced as dissonance or an error that is unwelcome~ In that spirit, the resistance to the introduction of new energies or frequencies often manifests in fears such as racism, religious discrimination, political or nationalistic separatism, to name a few instances~ These Energy Signatures are not entangled sentience~ These are dense, clouded, compacted awarenesses extremely limited in their ability for conscious evolution or change~

There is also density within densities~ For example, there are densities such as the Chicago Cubs and the Texas Rangers within the density cloud known as baseball~ Each team has its own Intellergy infused into that of the sport~ To take this further, the Intellergy of baseball as a sport attracts densities ranging from a child picking up that first bat and ball to the fan with his six-pack and pizza in front of the TV on a

Saturday afternoon~ For the most part, baseball's Intellergy is an embracing, welcoming sort, but others can become denser and more exclusive~ Religious, political, racial, sexist, and nationalistic belief systems fall into this category, where the true believers are willing to kill and die rather than compromise the structures of the densities in their Intellergy~ Manifestations such as war, genocide, racism, Jihads, and so on are some of the reactive forces generated when certain crystallized Energy Signatures encounter conflicting *frequencies* from another~ There is very little, if any, self-aware recognition of their state of entanglement with the One Mind, let alone _____~

For Blindsight, the idea of neuroplasticity is a simple way to stay in a state of play with the One Mind and _____~ By striving constantly for the state of the Signome, the mindful and self-aware entity can find ways to resist and dissolve the crystallization of their Intellergy~

For example, in the early stages of our understandings of the neuroplasticity opportunities of our own evolution, we observed that for our entire manifestation we had always put our pants on left leg first; our shoes and our socks left foot first; our shirts and coats left arm first~ The first few times we tried to put our pants on consciously and not reactively, with the right leg first, we practically fell~ To this day, it remains a conscious decision each morning which leg to choose, because if left to reactive consciousness, the left leg will always jump into first position~ The immediate *bounce* into present awareness from this simple act is always welcome and simply wonderful~

Recently, we noticed that in daily conversation we tend to pad many of our statements with the phrase, *you know*~ "I went to the store to, *you know*, pick up a few things." "The reason we do this is to, *you know*, get a better understanding." We have begun to observe that most often this practice is a function of rhythm or flow in our syntax~ Ultimately it is a completely unnecessary and unconscious evolution in our way of speaking~ At other times, the use of *you know* is a function of accumulated density in the form of insecurities, such as a lack of confidence in our ideas or our presentation~ There are occasions where the phrase is injected for emphasis or persuasion and actually acts to inject density into a conversation, relationship, or action~ Having observed this small nuance in our communication has fundamentally altered our awareness, as we now are more present and awake when we speak, thinking before choosing our syntax, and making conscious choices for this little Intellergetic reality we are forming~

The reader may see these as silly, insignificant examples, but for us it represents epiphanies reflective of hordes of compacted densities associated with virtually every action we take~ *Force of Habit* some may call it, but effectively we began to feel like a robot or one of those Gurdjieffian slaves that had been programmed to react blindly and move forward in our existence devoid of choice and input~ We began to notice this type of formulaic, derivative, and predictable behavior and reaction in virtually every aspect of our existence, as well as in others~ After decades of density accumulation, it has proven quite a challenging task to dissipate these *masses* and work toward a more participative, fluid, creative, and spontaneous process~

Next time you tie your shoes, try it backwards; put the left lace over the right instead of the other way around if that is your usual way~ Make your first loop with your less dominant hand and continue to finish the knot with opposite loops~ Observe the stimulation in your Intellergy as new harmonic centers begin to form~ Now, put your belt on upside down and observe the Intellergy there~ These and so many other smallest of actions are mere indicators of far greater robotic acts that draw us mindlessly through our enslaved manifestations, oblivious to all our potential~

It is this machinelike behavior, which is so predictable, that has empowered the global marketing experts to drill for all of the most personal of an Energy Signature's traits, selection cycles, buying tendencies, and likes and dislikes to build elaborate advertising strategies that are found in virtually every corner of the Terra Signature~ One can find the temptations for attractive density not only in the media, but with almost every click of a mouse on a search engine or walking down the aisle in a store or driving down the street~

To more broadly demonstrate human reactive, habitual, and robotic nature, we would offer two examples of how this behavior impacts the environment of the Terra Signature:

1) How many humans worldwide can you imagine leave the water running in the sink while they brush their teeth or shave? How many women and men shave while in the shower, using an even greater amount of arguably one of the world's most endangered resources? How many thousands of gallons of fresh water flow into our sewers and septic systems each minute of every day for these simple ablutions that should only require a few ounces in a glass or basin?

2) How many drivers sit mindlessly in their cars and trucks globally, every second of every day, at drive-through restaurants, banks, pharmacies, and the like using up literally millions of gallons of gasoline? How many carbon emissions are released? How much heat is expended into the ever-warming atmosphere? What advantage would there be if these people would simply park in the usually huge lot, take a few healthy steps, and increase the blood flow to their ever-expanding posteriors in their quests to super-size their wonder

meals, grab a stack of twenties, or pick up their choles-
terol medication?

Please choose from this time forward to observe these and
countless other robotic acts of the Human Disease that fill every
aspect of human existence, from the most mundane to the highest levels of cultural, political, and scientific endeavors~

If the reader would like to experience this phenomenon
immediately in another way, please perform the old act of
crossing your arms and hold the position for a few seconds~
Now reverse them, hold it, and notice the feeling of unfamil-
iarity~ The second position was a conscious choice~ The first
was a robotic response from a lifetime of repetition~ The older
the Energy Signature, the denser and more crystallized it be-
comes~ We hear the phrase, "They're so set in their ways" or
"That person is so hard-headed."

Please decide you never want that to be said of you~ We
most certainly do not wish it said of us~

Many human Energy Signatures are said to have *bad tem-
pers* or *stiff personalities*~ These all-too-common manifestations,
even including the dysfunctions of neuroses, sociopathy, and
the like are, in the end and in energetic terms, densely contracted Intellergy almost incapable of fluidity or creativity~ They
have become reactive, crystalized beings~

We would encourage the reader to begin bringing a spirit
of play to the idea of making conscious choices with respect to
such simple acts as using a fork or shaving with your opposite
hand instead of your dominant~ Wear your cell phone on the
right side one day and on the left the next~ Pay closer attention
to common verbal responses such as greetings or farewells, and
try to consciously improvise with a little creativity and more
presence in replies that are more reflective of the situation and
the person being addressed~

In simple ways, we believe one will be amazed at the rise in presence, awareness, and play that comes with uncomfortable choices made in the plainest of activities and situations~ These simple moments of enlightenment will lead to ever-greater awareness and participation in your manifestation~ It is the Blindsight position that this endeavor is very simply the only game in town if one truly wishes to consciously evolve and participate in the transition of the Great Shift~

Since these earlier times of which we speak, we have felt our awareness in regard to more and more of our daily routines begin to evolve and flex as our attention and intention make more conscious and present choices~ It has impacted everything from how we speak, how we drive, how we work, how we play, and how we interact with other entities~ It is a higher, more expanded frequency perspective that becomes a fun and wonderful game with practice, and we highly recommend it~

Gurdjieff told his followers over and over again to, "Remember yourself." If one believes this to be a simple practice, sit silently in a quiet, dark place, close your eyes, and just *watch* your brain at work for one minute — sort of like watching traffic go by at an intersection — as Alan Watts used to say~ Now, try to intervene and stop the flow of thought for ten seconds~ This would require no reaction whatsoever to any sensory response such as temperature, sound, observation, or idea~ There must be no thought of anything past, present, or future~ For most Energy Signatures in this highly accelerated and stimulating Information Age, this is extremely difficult if not virtually impossible at first~ It very quickly becomes very clear, especially when we attempt to extend this ten seconds for a longer period of time, that we are not actually the *thinker* in the event, but rather we are the manifested reactive *thought*~ We are effectively slaves to reactive responses, ingrained training, and experiential preconceptions~ When we encounter a stubborn, inflexible person, we might say, "He has a mind of his own." This actually couldn't be further from the truth~ In truth, this mind/brain has an Energy Signature of its own~

Gurdjieff would instruct his students to walk down the alley of one city block, challenging them to be aware of only their breathing and remember only themselves~ Rarely was a student able to report honestly doing this and never twice in a row~ Gautama, the Buddha, likened maintaining undisturbed self-awareness to "balancing a pin on a thread in a storm."

Blindsight would challenge the reader, as we do ourselves, to boldly play with a state of vigilance in your life and mind/brain, finding enjoyable ways to introduce more observation, flexibility, and fluidity into your thoughts and actions~

~~~*~~~

*A great way to accelerate this process is by playing a game we play every day and have played with our children since they were about three years of age~ The game is called **Searching for Signs of Life**~ It is really enjoyable and interesting in any type of crowd: grocery stores, department stores, special events, classrooms, congregations, walking down the street, and so forth~ It can be extremely powerful in one-on-one encounters~*

It only works if the player is present, self-aware, and in an open, receptive, nonreactive, observer state of mind~ We have found that maintaining deep and centered breathing with the slightest of smiles in the eyes and a sense of play in the mind makes this easier and a lot more entertaining~

While moving through the crowds of people, look impassively and fearlessly into the eyes of everyone you pass~ Be aware to not hold the contact too long or in such a way as to become threatening~ Soon one begins to observe that almost no one else around them is truly present or self-aware~ These entities are in a state of self-absorbed desire: I want~ I need~ I am hungry~ I am horny~ I am worried~ I am in a hurry~

I~Me~My~Mine~

*With practice, the game starts to take on certain elements of **mind-reading** as you will begin to see thoughts pass before their eyes like clouds obscuring their presence and their awareness~ It can be fun, especially with a child, to fantasize what it is these beings are desiring at that moment~*

Rarely, and this is something we haven't found in quite a long time, one will encounter a present, self-aware being in the midst of all the sleepwalking robots~ Our last brush with such a being was an elderly woman on a shopping scooter at a discount store~ Our connection was immediate and her countenance seemed to us to be almost glowing~ There was never any contact or conversation, and yet the exchange of Intellergy as our gazes met was tactile~ We could feel our Signome instantly enhanced, charged, connected, present, and entangled~ Then, with the slightest of smiles and a nod, she was gone, silently patrolling for her lowest-price guarantees~ This experience and the other rare and precious of its kind are at once inspiring and at the same time concerning that we don't see more present, self-aware beings~

~~~*~~~

There is a great movement currently within the Terra Signature taking advantage of the global Internet brain that is calling on all the so-called *Light Bearers* and other aspiring Signomes to not only merge and network their Intellergy, but also act as examples, teachers, and bridges for those who are beginning to awaken to their expanded potentials~ Many of these beings are direct channels to Qen5 intelligences who have been patiently waiting for this time and the Great Shift~ The highest profiles of such beings in recent history would be the likes of Edgar Cayce, Jane Roberts, and Judy Zebra Knight, all of whom were noted skeptics at their beginnings, only to find themselves channeling ancient and profound higher intelligences~ Although their skeptics are legion, their impact on global Intellergy is undeniable~ It is so naive, narrow, and arrogant to believe that our lowly human intelligence, self-aware only a few tens of thousands of years, is the pinnacle in a 13.7-billion-year-old Intellergetic Multiverse~ These older and entangled beings not bound by the densities and dualities of Qen4 Intellergy are always present and prepared to intercede for those who would quietly ask~ Throughout time, they have been called many names such as angels, spirit guides, gods, fairies, and more~ The names are only dualities and separations, but these intelligences are always willing to own them as bridges to the higher realms, ready for any to cross when they are ready~

The common Energy Signature is a reactive being, drawing to itself energy and information that resonates with its experiences, preconceptions, beliefs, fears, and desires~ It is an ultimately predictable, formulaic, and derivative existence with little expression of mindful creativity, choice, and action~ The irony lies in the insidious Human Disease characteristics that make the Intellergy completely unaware of their enslaved robotic qualities~

The goal of Blindsight is to awaken the curious Energy Signatures and motivate them to become more playful with the contractions and densities in their fields of Intellergy~ Strive to become more present in the NOW moment and observant of ingrained habits, reactions, discriminations, and fears that may have no other basis than past experiential density, with absolutely no relevance in present context~ We have no time or patience for those who would live from the old saying, "Once burned, twice shy." What is essential in these times is fearlessness and more of a "get back on the horse" mentality~

Why not?
Once more with feelin'~
— MWUSO

The attention and intention applied to our most mundane actions, such as our tending to our right leg going into our pants first instead of our left or brushing our teeth with our less dominant hand, grows over time into a more present and expanded awareness with regards to the larger situations, relationships, and decisions impacting our existence~ Imagine, for the sake of this Thojourn, a growing society of Energy Signatures with these higher awarenesses~ Can we conceive of Intellergies with the ever-present awareness that they are beings comprised of energy and consciousness entangled with infinite XPotentiality~ How different would our priorities be if we viewed our existence as energy-based Earthlings instead of Americans or Muslims, Attorneys or Farmers, Hispanics or Artists, Republicans or Socialists or others~ These labels represent endless progressions of collapsed wave functions, leading to more and more dualistic density in the Terra Signature~ They are the

basis for virtually all conflicts, whether personal, political, nationalistic, racial, or environmental~

Neuroplasticity is offered as one of the first steps toward
the greater understanding and eradication of the Human Disease, an opportunity to step outside of ourselves and discover
more expanded, present, and aware solutions to human dysfunction, self-destruction, misunderstanding, and conflict~

This chapter has approached this neuroplasticity subject
from many angles, but Blindsight would love the reader to take
away from it the spirits of play, creativity, adventure, fearlessness, watchfulness, presence, spontaneity, and fluidity that will
facilitate the expansion and receptiveness of the Energy Signature on its path to Signome and beyond~

NOTES

Latha by Erin Nicole

XXI

There is a Season

In physics, the Law of Conservation of Energy states that the total energy of an isolated system remains constant~ Energy can never be created or destroyed~ It merely transforms from one form to another~ For example, nuclear energy is transformed into kinetic energy, with the explosion of a nuclear bomb~ The amount of energy is unchanged~ Both these Intellergies are made up of energy and information, just as we are~

If we are sempiternal energy and consciousness, Intellergy, how can we possibly die? At our transition, we will never be individually our name, our body, our job, our family ever again~ However, the intelligent energy of the particular being is never lost; it simply is absorbed into the One Mind's *data field* as another way it has observed itself and enhanced its being-ness~ It may be helpful to remember that once you get past the dualistic reality of the One Mind, there is only one Intellergetic being~ This is what we have always been, since the Big Bang and the Inflationary Epoch~ As a result, having been *absorbed* into the One Mind, one's Intellergy is blended and interfaced, in multiversal terms, with the harmonics and resonances of the Multiverse~ These then coalesce into other manifestations~ These manifestations will, in turn, continue to absorb, store, process, and transmit energy and information on their path to further transition, observation, greater self-awareness, and evolution, eventually reaching another manifestation, leading to a transition point and another and another~ This is absolutely no different than the birth of a star and its inevitable collapse into a black hole~ No energy is lost, it merely changes form~ In this sense, our Energy Signature is the culmination of every single exchange of energy and information in the Multiverse since the Big Bang, and we continue to create and manifest in the Intention Stream~ As an Intellergetic Being, we are fashioned and exist as the entirety of intelligence, energy and awareness that is inherent in this stream~

~~~*~~~

*As a sidebar, the Theosophical Society — self-designated as **an unsectarian body of seekers after Truth who endeavor to promote Brotherhood and strive to serve humanity** — came into being just before the turn of the twentieth century~ They were an interesting, if not controversial, group of **seekers** that some on the path to dimensional transition might find illuminating, but for the purposes of this particular diversion we would like to refer to the theosophist's belief in something they called the **Akashic Records**~ These, they believed, are a compendium of all earthly human thoughts, experiences, desires, and actions contained in what H. P. Blavatsky termed "indestructible tablets of the astral light." Obviously, she and most of the other members of the society were openly occultists~ As most occultists, they were extremely gifted at pointing the finger at the moon of the Multiverse, but they nevertheless coined some thought-provoking imagery along the way~ Alfred P. Sinnett, a student of what he* *called **Esoteric Buddhism,** wrote of a Buddhist belief in "a permanency of records in the Akasha" and "the potential capacity of man to read the same."*

~~~*~~~

For Blindsight, this is simply another way of looking at the potential that awaits a fully expanded and entangled Signome in immersion with One Mind and _____~ All of this so-called Akashic Intellergy exists as a sort of Multiversal *library* that is accessed through an Intentional search by a fully expanded Intellergetic Being~ It is the Blindsight position that by the time such an entity has attained complete immersion into _____, there will be little concern for or interest in the records of human existence; however, along the way, these concepts can be helpful as material for Thojourn explorations~

The Tibetan Book of the Dead charts the journey of the soul from the moment of death through to reincarnation or, for some, the passage to higher, more expanded space~ The book is assembled from the testimonies of Lamas over centuries who, after meditating since childhood, were supposedly able to retain self-awareness throughout their transition and rebirth~ The current, living Dalai Lama is said to be one of these beings~ Along this journey, the soul apparently passes through various realms of turbulence, conflict, and even torture~ It is then given the opportunity to evolve into a higher being~ At this juncture, the Lamas speak of the final *test* being vivid visions of every possible form of sensual congress, combined with overwhelming forces of sexual attraction and urges~ Most are irresistibly drawn back onto the wheel of the Karmic Cycle to work through unfinished challenges~ In supposedly the case of the Dalai Lama and others, a conscious choice is made to return as an example and teacher, facilitating the evolution of others~ It is only the purest and fully realized beings that are able to resist this ultimate trial~ They are welcomed into a realm of spirit energy, a heaven of sorts, a Qen5 space of infinitely expanded creative potential, where whole new sets of evolutionary trials await on the ever-evolving path to oneness with _____~

There is a parable of sorts relative to the discussion of reincarnation that is attributed to the current Dalai Lama, who speaks of death like the shattering of a china cup filled with tea~ One imagines many broken pieces of the cup strewn around the surface, while the tea separates into pools and drops~ One can never put the cup or the tea back together as they were~ In Tibetan Buddhism, the body of the deceased can never be restored, but the Intellergy and the place along the path to Enlightenment remains intact~

Once again, Blindsight would prefer to look at this cyclic process of transition more energetically~ In this, our Multiverse, the One Mind is fueled by the original intention of _____ that led to the emergence from the Singularity of the Big Bang into the Inflationary Epoch; Light~dark, male~female, AC~DC current, hot~cold, in~out from gravitational sources, the dual properties of light, mass~energy, God~human, etc.~ Duality stimulates creation and destruction~ Densities of Intellergy, along with their reactive fluctuations, fuel the energies of tension and release, necessity and desire~ It becomes the irresistible force, the immovable object~

The Circles of Life are the direct result of the ebb and flow of Intellergy's dual nature~ Soon after birth, within usually five to ten years of age, the early Energy Signatures become aware of their mortality, realizing they will eventually transition, or in their Qen4 reality, die~ This realization, from that moment on, subliminally underlies virtually every thought, word, and deed~ For some it is the drive to live life to the fullest, whatever that may be for them~ For another, it is the will to be the best, the fastest, or the smartest~ For others, it is the freedom to indulge in every appetite and desire~ One might hear them say something like, "Why eat healthy and work out? I could get hit by a bus tomorrow."

After the passing of a great pleasure or at junctures of pain, loss or frustration, there is always a quiet reminder that this too will pass~

Yogis teach that in the quiet and stillness between every breath, at the peak of every toss of a ball, the silence between each note of the symphony, there is emptiness and space~ This is where the XPotential for the possibility of immersion into _____ resides~ This is the opportunity for the Energy Signature to become Signome~ With practice and disciplined intention, the Signome can remain in this state indefinitely or even permanently~ It is from this space that many have consciously chosen to leave their Qen4 manifestations and move into Qen5 and higher realms~ Numerous gurus, yogis, shamans, mystics, and prophets have supposedly chosen this path~ Jesus and the Buddha are widely believed to have done this~ You who are diligent in your intention to explore some of the thought journeys such as those put forth in the pages below or some of your own may find your awareness at one with the *ocean* of fully expanded space, as consciousness without an object, Nirvana, Samadhi, or Satori~ As a being fully entangled and realized as its true nature, _____, many Signomes discover they have no desire to return to the turbulence of Qen4 Intellergy manifestation~

Sola Lochan by Erin Nicole

XXII

God?

~~~*~~~

*There are many versions of an ancient parable the Sufis call **The Blind Men and the Elephant**~*

One version of the story goes that three blind men set out in search of a great and mystical creature known as an elephant~ For centuries, prophets had taught that finding and knowing the elephant would lead to infinite knowledge, understanding, and enlightenment~

After a long and arduous journey, they at long last came upon the astonishing beast~

The first blind man to reach the Elephant ran into the leg of the animal and, with tears running down his face, began to exclaim, "I have found the Elephant and he is a tree!" He immediately turned and ventured off into the world, telling all who would listen of the knowledge and gifts of spirit he had found in his experience with the Great Tree that was the Elephant~

The second blind man very soon after came into contact with the tusk of the Elephant and he turned with amazing joy, calling out, "I have met the Elephant and he is a plowshare!" He then turned and struck out in a totally different direction, proclaiming to all of the spiritual awakening he had experienced with the Wondrous Plowshare that was the Elephant~

The third blind man, by this time, had also found the Elephant and grabbed his tail, shouting, "The Great Elephant is a rope! He is a rope!" He too ran into the world proclaiming his enlightenment in knowing the Incredible Rope that was the Elephant~

Very soon each of the three blind men found that they each had a group of followers who were hungry for the message of their enlightenment experience and began to mold their lives based on their every word~ They wrote down their teachings and put them into books so that others might read, study, and find the same happiness, inspiration, and spiritual wealth they had found~

They constructed great houses where the multitudes could gather and be in the presence of the spirit of the Elephant~ Extraordinary temples of worship, institutions, and schools were established to examine and interpret the teachings of the great blind prophets~ Smaller groups of the wisest scholars began to meet, endeavoring to further dissect the writings of their respective faiths~ They continually found nuances, translations, and interpretations in the texts, leading to cycles of ever-deepening discussions~ Eventually certain factions began to break away as they found their elucidations of the teachings to be in conflict with others who had different understandings and versions of the very same words~ These traditions, along with various newly constructed books, began to condense into more compacted states that became perhaps more manageable and defined~

Descendants of these original followers began to take the larger number of books, condensing them into just one book that they may have titled something like a "Bible," a "Torah," or a "Koran." All of them would be speaking of only a certain, singular aspect of the one elephant: the rope, the plowshare, or the tree~ A prospect of confusion and conflict would be unavoidable~

Time passed, as did the lives of the three blind men, and each of these schools of thought evolved, more and more convinced that theirs was the only true knowledge of the Elephant~ They began to compress energetically into a cultural essence of exclusivity, where if one did not believe as they did, they would have no possibility of ever knowing the true spirit of the Elephant~

Each of the schools began to feel increasingly threatened by the conflicting philosophies, and they grew aggressive in their proselytizing and the defense of their beliefs~ This schism inevitably led to confrontations and eventually wars~

In the end, sadly no one had ever truly understood the true nature of the Elephant~

~~~*~~~

Blindsight would offer up the Elephant in this story as _____ and, for the sake of discussion, the three blind men could very well be the roots of Judaism, Christianity, and Islam~ All of these doctrines believe in one God~ All of them stem from the experiences of one man, Abraham, and yet the modern *crystallization* and exclusivities of their disparate philosophies have led to some of the most violent confrontations and attacks~ One of the greatest threats to the Terra Signature in the millennium is planetary destruction as a result of conflict over _____~

If there is a more absurd premise, we would love to know it~

The Merriam-Webster Dictionary defines religion as the service or worship of God or the supernatural: commitment or devotion to religious faith or observance~ They go on to define it as a personal set or institutionalized system of religious attitudes, beliefs, and practices~ The archaic reference is *scrupulous conformity*~

Many of our early readers have passionately challenged Blindsight as an attempt at crafting some kind of new religion, apparently for our repeated references to the indescribable, unnamable source of all that we simply call _____ ~ If we were to work within the context of the definition of a religion, this would imply that Blindsight is suggesting some sort of system of beliefs to be held with ardor and faith~

One of the core tenets of this work is the transcendence of all belief systems, with the understanding that nothing but humble acceptance of one's true nature is required for immersion into the eternal presence of _____~ Blindsight offers merely a medium for one's unencumbered exploration and enhancement based on truths revealed in millennial observations~

If there is any correlation between Blindsight and religion, it would be found somewhere in the billions of humans in the Terra Signature for whom ancient religious texts, crafted by uneducated, untrained observers with pastoral backgrounds, seem to be struggling for relevance in millennium Intellergy~ If any of these seekers would find some alternative for their quest herein, then it is our sincere hope our work is taken not as religion or belief structure, but rather as a reflection of their own true nature~ Similarly, if one in any way takes the language of

Blindsight as the pretense of some sort of *scripture,* please refer to the repeated intention that language is ultimately illusory, and serves no purpose other than the *finger pointing at the moon~*

These statements should never be taken as any affront to these amazing works, but it has become sadly apparent that the *leaps of faith* required in these books for true entanglement with _____ continually engender dualistic relationships that are ultimately illusory~

Blindsight recognizes the belief that God, Brahman, Atman, Allah, and any other words for a supreme being represents some sort of dualistic relationship between the believer and this being~ One finds believers in the act of prayer with their hands held in the air, reaching for a connection with their god~ We see athletes kneeling or pointing a finger into air in thanks to their deity of choice for their most recent goal or victory~ Each day, millions halt their day several times to engage in the ritual of praying to the east, as if some representation of this being is there waiting to hear their prayers and absorb their belief~

Many of these Intellergies will also expand this dualistic relationship to include belief in destinations like heavens that would offer everything from wings, harps, golden streets, and gates of pearl to hordes of virgins~ All of these wonderful and supposedly desirable rewards are allegedly to be made real at the time of death, while still maintaining one's earthly identity, personality, and body in spirit form~

At the same time, existing as an atheist in complete denial of something greater than one's Energy Signature is also completely irrational within the context of anthropopathism and the One Mind~ An atheist lives in a very small, contracted Intellergy, and often finds it difficult to take advantage of the infinite Xpotential of creativity and existence that awaits with immersion into higher Intellergy~

To go on and on about a supreme being is absurd~ Please simply attempt to manifest from the awareness that _____ is~

All objects, thoughts or words that endeavor to describe, name, or worship _____ represent collapsed wave functions and create dualities that are illusory~

Remember, as a fully entangled Intellergetic Being, you are in the constant presence of _____~

You are eternal, omniscient, and omnipotent~

All that is required of you is to accept it as your true nature~

_____ is~

NOTES

Machdnach Photo by G~

XXIII

The Qen Trinity

We are _____~ According to Genesis 1:27 in the Christian Bible, we are "created in God's own image." When we know and act with mindful intention from immersion in _____, we are the creators of our reality~ When we are not present and self-aware, we are merely unconscious reactions and densities swept along in the runaway flow of the Qentinuum~

At this point in the simulation, we sincerely believe that if the reader has remained with us up to now, hopefully you have begun to imagine yourself as Intelligent Energy living from a core-level knowingness of infinite potential energy, consciousness, and present self-awareness; Xpotential existence as Signome~

From this perspective, try to play with answering two questions:

What existed before the Big Bang?

If the Multiverse is expanding, then into what is it expanding?

It does not seem like too great of a stretch to imagine that had we spent most of the resources of science and philosophy mentioned earlier in pursuit of these answers, many of the questions that now bewilder and challenge us would possibly have become clearer~ If we could understand even with limited certainty what energies, forces, and consciousness went into the creation and formation of this Multiverse, most of the mysteries that confound us today would most likely vanish~

One very interesting conversation out there today begins in the mind of Oxford University philosopher Nick Bostrom~ He, along with astrophysicist Neil deGrasse Tyson and others, have postulated some very compelling mathematical models that would have the Multiverse existing as holographic construct or matrix-like simulation, much like one of our online video game universes~ They envision a planetary mass computer able to simulate the entire history of humankind, including its absorbed psychology, bio-technology, quantum understanding, mental/emotional constructs and yes, even buyer tendencies~ They call this ancestor simulation, and such a computer theoretically could create such a simulation in less than one-millionth of one second~ They also project that a pre-human, meta-intelligent civilization would be capable of reproducing an astronomical number of these computers~

We have an acquaintance who has been an online Star Wars gamer for over a decade~ In his Qen4 manifestation, he is a quiet, shy, and unassuming man working as a sales associate at a home entertainment showroom~ However, for several hours each night and at times all night, he becomes one of the highest-ranking generals in the game's rebel army~ He has amassed many businesses, armies, and supply caches~ Other armies join with him each night because of his long-standing record of victories~

In this alternate Multiverse, this unobtrusive individual becomes a confident, aggressive leader who is more and more compelled to remain in this world he has created rather than return to the relative impotence of his Qen4 existence~

This simulated Multiverse is a computer-generated composition of ones and zeroes~ When viewed Intellergetically, it becomes a manifestation of intelligent energy activated by intention~ From a One Mind's perspective, this simulation is no less real as an aperture for observation than our friend's Qen4 existence~

When he is fully engaged in the game, his Intellergetic identity, his Energy Signature, has expanded~ He is at that time involved in the process of collapsing wave form after wave form to advance his positions of victory, power, and control that ultimately have values that differ only in frequency from the daily successes in the life of a humble salesman~

Supercomputers will achieve one human brain capacity by 2010, and personal computers will do so by about 2020~ By the 2030s, the non-biological portion of our intelligence will dominate~
— Ray Kurzweil

ETEWAF = Everything That Ever Was Available Forever
— Patton Oswalt

Although most of the Quantum community immediately rejects this concept, none have been able to effectively disprove the math~ In the film *Avatar*, the avatars in a simulated universe were self-aware and capable of choice~ Is it possible that we are no different? Could there be some adolescent being with intelligence as far beyond our own as we would be to an insect who is manipulating our cosmos to collect points or defeat a competing player?

Is our Intellergy simply a construct for such a simulation? It is unlikely that we could ever know~ Blindsight allows that simply because of the anthropomorphic nature of our Multiverse, simulated or not, we as Energy Signatures are still capable

of choice, free will, and intentive immersion into _____~
Behind even the most advanced Meta-intelligence, up to and including One Mind, _____ is~

Could it be equally feasible that our universe is merely a reaction to a larger event, like a solar system absorbed into a black hole only to emerge out of a *white hole* as an entirely new universe?

These little Thojourns spawn many others and there obviously are no rules~

~~~*~~~

It is probably relevant at this point to introduce a disclaimer and remind anyone who would include any of these Mind Experiments of ours into their belief systems or construe them as our own personal constructs of reality is completely missing the point of the entire dialogue~

~~~*~~~

The questions still remain concerning the source, the destination of our existence, and our ability to know it.

Before the Big Bang, there existed *(and presently exists surrounding our Multiverse)* an immeasurable, sempiternal *field* of consciousness with no objects, yet with unlimited creative potential; an unformed *suchness* to use one of the Buddha's descriptions~

Infused within this field was *(and is)* infinite, limitless energy, also lying dormant as endless potential~

Please take a short moment (longer is recommended) to ponder the significance of the concept of Limitless Potential with regards to Energy and Consciousness~

This is an oversimplification of an idea construct that is unknowable and completely indescribable without assigning object, imagery, language, or energy, and yet here we are~

This field of consciousness/energy we call Qen~ It would and will lay dormant for all infinity~ It responds to only one thing: *INTENTION~*

_____ is the source of all intention~

_____ is infused into every aspect of every Multiverse, therefore the Multiverse must be self-aware~ From the dawn of human belief in a higher power, Energy Signatures have assigned a non-human supreme being and lesser ones like gods, spirits, demons, and the like human traits, emotions, and intentions~ This is known as anthropomorphism or, in the case of its application to deities, anthropopathism~

For example, God is often depicted as an old man with thick, swept-back hair and a great beard, but so was Noah, Abraham, Moses, Zeus, Poseidon, Odin, the North Wind, and Santa Claus, to name a few~

The Greek philosopher Xenophanes argued against anthropopathic conception, saying, "the greatest god" resembles man "neither in form nor in mind."

If our Energy Signature was not infused with _____ consciousness, we would not be able to know we are present, alive, or mortal~ This might loosely compare to the idea of Vishnu, *the one who enters everywhere,* in the Vedic tradition of the Hindus~

From the above premises we would like to posit a Blind-sight Thojourn we will call the *Qen Trinity~*

There are three fundamental aspects of this trinity:

1. _____
2. *Energy*
3. *Consciousness*

Blindsight loosely compares these three elements to the Christian Father, Son, and Holy Spirit, or the Hindu Shiva, Brahman, Vishnu, or perhaps the Chinese Taoism, Confucianism, Buddhism~ However, it is the Blindsight position that this Qen Trinity is the *Trinity* from which all others emerge~

In this particular game, the infinite, all-knowing being-ness that is _____ becomes inspired and is, at the same time, maybe a little lonely and bored~ At this point, *INTENTION* is *sent* into Qen, the infinite unformed possibilities of the consciousness field and limitless potential lying in wait within the undisturbed energy pool~

A Multiverse is born~

BANG!! (A big one!)

Impregnated with the *INTENTION* of _____, this Multiverse proceeds to expand and evolve from its Inflationary Epoch~ This new beginning continues to stimulate creation of infinite Multiverses and dimensions infused with self-aware intelligence to examine, observe, and reflect upon itself using the dualities of Intellergy as fuel for further creation/observation cycles that will endlessly repeat~ _____ may appear a little vain, or maybe a little childlike in this game, but *NO RULES,* remember?

There is also Bostrum's meta-intelligent simulator possibly between us and _____ ~

Acclaimed black-hole physicist Stephen Hawking came up with some math implying that nothing existed before the Big Bang~ However, he was able to show that some sort of spatial potential could occur even in nonexistence~ Hawking theorized that under *extremely bizarre conditions,* empty space can transform into time, and time can morph into space~ His math projected that if the tiniest bit of space turned into time, the mere appearance of time would trigger a Big Bang because motion, cause, effect, and energy would occur as a result~ He came to this conclusion because the Big Bang was an event where there was *nothing* and all of a sudden, there was *something* with no observable precursor or cause~

Good times for Mr. Hawking; however, Blindsight would still offer that time, space, or spatial potential still had to have existed for his mathematics to work, along with the intentive intelligence and energy to activate them~

In ancient Navajo tradition, there is a Creation Myth called *Dine' Bahane'* or the Story of the People~ It requires nine nights to recite~ It must be memorized, and it entails years of practice and study for an initiate shaman to master~ It has never been accurately transcribed or written down~ It is chanted in a high, quavering pitch with a very unusual five-tone scale that most often moves even the most resistant of nonbelievers~

For us, this partial, quite wanting translation offers another perspective of Intention's role in creation:

> *Of it he is thinking, he is thinking~*
> *Long ago of it, he is thinking~*
> *Of how darkness will come into being, he is thinking~*
> *Of how Earth will come into being, he is thinking~*
> *Of how blue sky will come into being, he is thinking~*
> *Of how yellow dawn will come into being, he is thinking~*
> *Of how evening twilight will come into being, he is thinking~*
> *Of dark moss dew, he is thinking~*
> *Of horses, he is thinking~*
> *Of order, he is thinking~*

Of how everything will increase without decreasing, he is thinking~

He for the Navajo is the Creator or _____~ The Navajo saw _____ as a supreme being outside of themselves and yet they intuitively understood that intention was and is the source of creation~ How did _____ come to exist? For Blindsight, that will have to remain a question for another time~

In Matthew 18:3, Jesus was quoted as saying, "Truly, I say to you, unless you turn and become like children, you will never enter the Kingdom of Heaven."

As Blindsight continues to emphasize the amazing gifts, advantages, and potential available to an Energy Signature growing into a spirit of play, isn't it a compelling scenario to imagine _____ as the ultimate playmate?

The tiny, insignificant Energy Signature writing this narrative is in no way material~ As mentioned earlier, we are not sacks of meat with bones, organs, and blood~ We are fleeting *blooms* of energy, information, and awareness with a mind/brain concentration of density that behaves much like the concept of a computer chip~ We are Intellergy~ In this respect, this Intellergy draws in information and energy, stores and processes it, then transmits~ Even an amoeba, a rock, or a hydrogen atom performs the same functions ad infinitum~

This is _____ at play in the Quantum Trinity~
Signome~

NOTES

Mir by Erin Nicole

XXIV

Turn, Turn, Turn

This is a chapter on true nature~ In the end, our true nature is _____, but our discussion is more about the cyclical path of the Energy Signature that is taken on its way to immersion in fully expanded space~ Please understand that the *space* of which we speak is anything but an empty void~ Fully expanded space is an infinite medium of XPotential intelligence and energy awaiting intention, and at the same time completely and comfortably at rest as undisturbed possibility~

The Blindsight course of travel is quite different than that of *The Tibetan Book of the Dead* in that it doesn't involve all the turbulence and torture, thank goodness; unless that is a consequence of your intentions~

Our path would begin with an Energy Signature's first awakenings to its ability for conscious evolution and the overcoming of the Human Disease through the dissipation of accumulated densities involving experience, expectation, preconception, and belief~ What begins at this point is a journey of possibility and potentiality of an infinite nature~ By boldly embarking on this journey, Intellergetic Beings will most often experience the initial immersion awareness that is Nirvana, Samadhi, or Satori, even if only for very brief amounts of time~

Once a being has found themselves in this immersion into _____, their true nature is revealed and from this point onward no experience will ever be adequate~ The being is locked like a ship following a homing beacon in a storm, and there is an inevitability that to deter from the course would mean certain death and destruction~

In this context for Blindsight, the first step on the path is to **SEEK** your true nature~ Once you have experienced even briefly fully-expanded, self-aware space, either as we have through life experience or through meditation, epiphany, or hallucinogens, the journey has begun and like the beacon, it will lead the seeker on a path of continual choices and efforts fraught with enlightenment, disappointment, boundless joy, and diversion~

~~~*~~~

Once early in our awakening, we were driving on a dark, and fortunately, empty country road in the middle of the night, pondering these ideas and following the flow of our reverie into ever higher realms of thought~ In an instant we found ourselves in a state of Nirvana, completely separated from our body~ We returned to our body, not knowing how long we had been away, about one hundred yards out in a cotton field~ We were unharmed, as was our car, but it was some time before we felt able to gather enough control of our physical body to drive ourselves out of the field and home~

The next day, we remember going to find one of our best friends and enthusiastically told them, "I saw GOD last night!" They had a complete look of "How am I supposed to respond to something like that?" on their face~

Since that time until this writing, we have been extremely careful with whom we share these types of experiences~

*We have also learned not to operate heavy machinery
under the influence of _____~*

~~~*~~~

We have related this story to let the reader know that if one chooses to seek this experience, it can come upon you at any time— waking or sleeping, at work or play~

The second step on our journey is to **KNOW** our true nature~ To know something is to perceive it directly, regarding it as truth beyond doubt with complete certainty and clarity~

Once an Energy Signature comes into contact with their true nature of immersion into _____, the challenge becomes to find and develop ways to stay in this state for longer and longer periods of time; to learn to **KNOW** these states~ For most, this will require experimentation as to what is the best method or practice that will resonate and harmonize with Intellergy's existing frequencies in this particular manifestation~ Some form of meditation or creative visualization in conjunction with breathing exercises and playfulness is usually a good place to begin~

In combination with these practices, we have found that reading, studying, and absorbing the words of the great beings who have gone before tended to light the path and, in some ways, offer *signposts, detours, and pitfalls* along the course~ Some of those who inspired us include the Buddha, Aurobindo, Jesus, Yogananda, John C. Lilly, Shankara, Alan Watts, Ram Dass, Black Elk, Franklin Merrell-Wolff, G. I. Gurjieff, Shankara, and Thich Nhat Hanh~ These teachers had the most impact for us because we were able to read and absorb their actual words rather than secondhand reports from untrained and non-realized observers~ Perhaps they can also assist the reader, although there are many others to be found~ We would encourage those who would embark to trust that they will find higher Intellergies continually attracting them along their course~

Once an Energy Signature is able to remain in the state of Nirvana for an extended period, they are Signome: a fully en-tangled, self-aware, and present being at one with _____~ They *KNOW* their true nature~

In this fully-expanded and aware state, the Signome is now a being of unconditional *LOVE*~ There is no duality~ There can be no misunderstanding, no competition, no jealousy, no fear, nothing to be lost or gained~ There is no sense of body, mind, personality, or relationship~ The Signome is embodied *Source*~ There is no cognizance that this is a destination or culmination~ It is a simple, present beingness~ There is absolutely no sense of urgency or attraction for the fully-expanded Signome that would compel a return to Intellergy form~ For many, there is no choice~ There is only the clear and simple realization that one is now *home*, infused in _____~

For most beings, the Signome experience still contains residues of momentum and resonance generated from Qen4 densities that have not completely resolved and dissipated~ These densities hold charges and unresolved tensions in duali-ty form that coerce the being to return to Energy Signature form and begin the journey once more~

This is the step Blindsight calls *LIVE*~

Intellergy must now begin their journey to *LIVE* and search once more for their now-evolved true nature~ This is likened to a sort of Intellergetic reincarnation~ Self-awareness has altered now, because the Energy Signature has acquired or absorbed higher, more expanded harmonic frequencies of energy and intelligence~ Intellergy has consciously evolved~ It follows that an Energy Signature may now find itself more aware of certain opportunities, impressions, or teachers along the path this time around that may have been overlooked before~ It is clear from our personal experience that the vividness and the memory retention of our out-of-body, near-death experiences and meditations continue to breed familiarity with and attraction to higher states~ Certainly each time we come back from a place of expanded awareness, it becomes easier to return and remain~

Within the **LIVE** cycle now transitions the inescapable drive to **SEEK** once more renewed transition points and Intellergy on the Qensional Bridge~

With diligence and intention, every newly rejuvenated and enhanced Energy Signature along the cycle should be able to advance more quickly along its journey~ At the juncture of each plateau of **SEEK, KNOW, LOVE,** and **LIVE,** more and more confidence and understanding of one's true nature emerges~

The destination, after all, is no destination at all~ We are always present in a state of oneness with _____~

All we must do is unconditionally love who and where we truly are now~

Be here now~
— Ram Dass

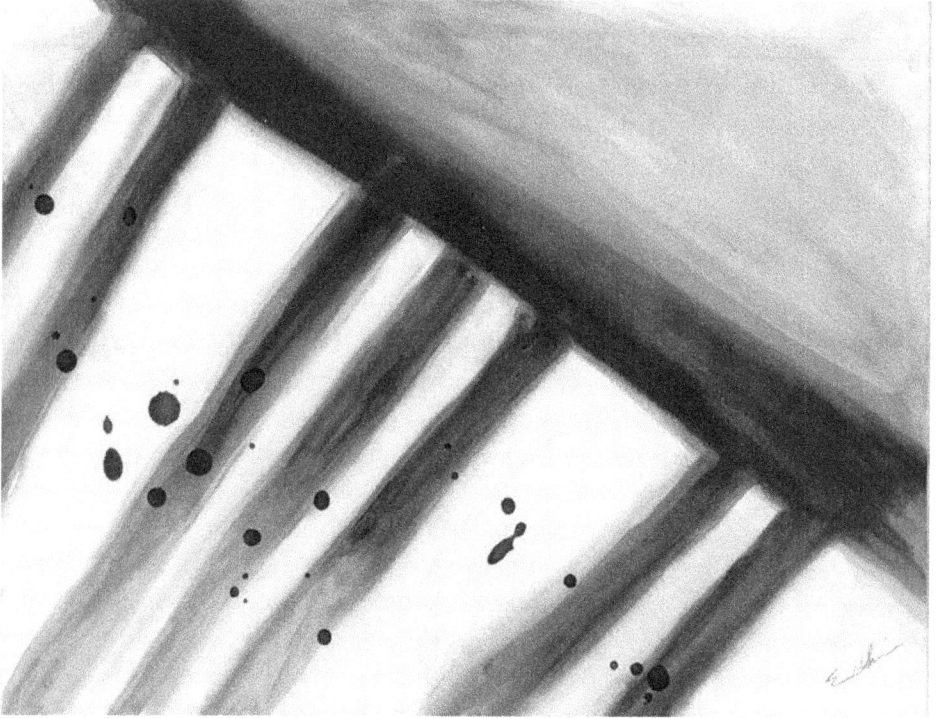

Lasan by Erin Nicole

XXV

A Dark Matter

We should enter a disclaimer of sorts at this point because we have struggled with the desire, as well as the necessity, for leaving this chapter out of this simulation~ Our overall intention has been and remains to leave the reader with positive inspiration, challenge, hope, and intentive purpose~ While we have endeavored to seed this following section with these elements, the overall is that of extremely negative Intellergy, which has gained far too much power and influence in this age of the Great Shift approaching the Quantum Point~ "It's better the Devil you know than the Devil you don't" is what our Grandfather would say~

A lie can travel halfway around the world
while the truth is putting on its shoes~
— Anonymous

The above quote has been attributed to everyone from Mark Twain to Thomas Jefferson, finally settling on Anonymous, but from whomever it originated, the Intellergy it expounds rings true~ The Human Disease is based on misconception, illusion, and diversion~ It is ravaging among us in the millennium, especially in the political, religious, media, and business arenas, complicated by the unrelenting spread of social media, so we find it impossible to omit~ Please read the following boldly with every confidence that you have all the infinite powers of intelligence and energy at your disposal for defense against these forces~

With regards to physics and cosmology, dark matter is a phenomenon that has, of this writing, not yet been identified by any type of observable process, and yet it apparently makes up about 27 percent of the energy and supposed mass in our observable universe~ It has been given the name *dark,* as it does not emit or even interact with electromagnetic radiation, including light, so as to be undetectable by current imaging technologies~ It is invisible to the entire electromagnetic spectrum~ As a result of dark matter, measurements of the apparent masses of large-scale structures, such as galaxies, are considerably greater than masses measured of the observable *luminous* matter~ This indiscernible mass has dynamic influence on gravity, which extrapolates into light propagation, formation and destruction of stars, and ultimately into human physical realities~

~~~*~~~

Always keep in mind that mass, gravitational influence, and physical realities are illusory in a Multiverse, comprised entirely of energy and intelligent information~

~~~*~~~

About 96 percent of our universe is made up of the combination of dark matter and dark energy, and as of this writing physicists really have no idea what they are~ Blindsight would like to remind us to always be looking from the outside inward, holding the context that we are also made up of 96 percent dark matter and dark energy~

What then are the implications for an Intellergetic Being?

Up to this point, most of our narrative has tended toward positivity, potential, and possibility~ However, it is important that we do not diminish the power of destructive, entropic forces in the Multiverse that would seek to feed the Human Disease and thwart efforts to expand Intellergy and immersion into _____ ~

For Blindsight, it is the Qen4 densities, dualities, and attachments of the Human Disease that impede expansion and connectivity~ There are Intellergies in the forms of *dark* beings, entities, and forces at every level of Intellergetic evolution, with highly intentive missions~ These beings are empowered with Intellergy Forming abilities intentively creating language, imagery, scenarios, belief systems, and objects crafted to distract, misdirect, and addict Energy signatures~

In Christian mythology, the fallen angel, Lucifer, had lived with God and was known to possess great beauty and piety, but he was thrown down because of his extreme pride and treacherous overconfidence~ As the ultimate I~Me~My~Mine embodiment, this devil entity could be considered the architect of the Human Disease~ Legend, fantasy, and myth would have us believe in Satan's Legions~ In an Intellergetic sense, this not an unfair metaphor, with the accelerated density of illusory-based perception filling the Terra Signature as we approach the Quantum Point of the Great Shift~

For Blindsight, there is no fear, as these dense beings have simply not been awakened to the Intellergy of their true nature~ Ultimately, they wield absolutely no power that is not afforded them by an Energy Signature with density accumulations of anxiety, fear, attachment, belief systems, and preconception~

Every intention and result of these dark Intellergies is destined for entropy and failure~ _____ simply cannot absorb this darkness in that form~ However, also keep in mind that the dark energies could never have existed without the initial intention of _____~ It is from _____ that they have emerged, and *filtered* through One Mind, providing simply one more aperture through which _____ observes itself~

> *When you look at the dark side,*
> *careful you must be . . .*
> *For the dark side looks back~*
> — Yoda

It is interesting to note that Multiversal Simulation Theory becomes a little more compelling if one imagines our Intellergies existing in this monstrous *video game,* where forces of good and evil, creation, and destruction wage for domination, leading to Armageddons, with victors wielding ultimate power~

The uncomfortable truth is that the Terra Signature nearing its Quantum Point of the Great Shift has an extraordinary density of this dark Intellergy~ The Human Disease has contributed to compacted awareness in virtually every area of reality, from religion to politics, socioeconomics to environmental, global to personal~

In game simulation, both forces are made available of an arsenal of weaponry with which they wage their respective battles~ With regards to our own Intellergetic conflict, the struggle is fought on the fronts of race, religion, culture, nationality, and sex~ With a few simple *keystrokes* applied with determined, repeated, and organized intention, these dark Intellergies disseminate bias, hate, confusion, and lies and are able to incite protests, alter legislation, influence elections, inspire terrorism, and ruin lives~

Witness the explosion of wave upon wave of anti-police Intellergy in social media as a result of a few too many tragic miscalculations by officers~ What has resulted is the escalation of attacks on officers~ The good cops who always courageously go on the job every day feeling a target on their backs now sense a homing device attached to the bullseye~

In elementary, middle, and high schools around the world, cyber-bullies find the small, the weak, or the insecure, honing in on them and spreading untruths or threats~ These actions have led to unprecedented rises in depression, suicides, and violent reprisals~

Each day on websites and in social media, Christian and Muslim followers launch provocative, fear-based messages inflaming the ever-broadening chasm between two philosophies that ultimately have quite a lot in common at the end of the day~

The examples are myriad, but the weapon that is employed is simple: *Duality~*

Divide and conquer: such an unassuming strategy when all of us with the Human Disease have already bought into the structure~ With the perspective of Quantum Entanglement and Superposition, all these differences, discrepancies, and misconceptions harmlessly fall away~ Nevertheless, the war wages on~

There is literally an *army* of dark energy *criminals* at large on the Terra Signature at this time, exerting every intention in their considerable powers to shift the Quantum Point in their favor, away from the intentions of the Light Bearers~ Please remember that to the impartial Observer and by association, One Mind, either consequence is acceptable~

The outcome for Blindsight in the Great Shift must be one of expanded, aware consciousness with infinite potential~ Nevertheless, these criminals exist, and many of them are in high-profile Energy Signatures as political, business, media, or religious leaders~ Many also manifest as extremely successful artists, actors, and musicians~ With a modicum of presence, observation, and awareness, these beings are easily recognized and dismissed simply by virtue of their works~ However, there survives an insidious element within this dark Intellergy stream~

For example, on the fringes and in the nadirs of the Internet we find what is being called the *Dark Web* or the *Deep Net,* where the worst of the Intellergy Criminals network, conduct business, and power agendas of the most nefarious natures while building extremely predatory systems and relationships~ With ever-increasing security firewalls and cleverly masked presentations, these shadowy, sinister depths are expanding at almost exponential rates~ Law enforcement's best are overwhelmed and understaffed as they attempt to track, trace, and combat a kaleidoscopic beast~

At this point, we will enter our strongest word of caution for any emerging Energy Signature on an intentive path to transitional harmonics~ When diligent work begins to dissolve and disperse densities and accumulations of derivative Intellergy, the resonances of not only the Energy Signature but also the individual centers begin an alteration in terms of frequency as they become more fluid and receptive to more expanded timbres~ These transformations will engender feelings of profound joy and inspiration for continuation on the path of transition to Qen5 space~ At the same time, like blood in the water to a Great White Shark, Intellergy Criminals sense and feed upon these frequencies of openness and newfound hungers for enlightenment that are transmitted by the newly awakened Energy Signature~ In the following chapter rendering, the reader will find a number of examples of how such dark Intellergy can insinuate itself into the newly expanded energy centers of an awakening Intellergetic being~

At the same time that these Qen4 monsters move among us, there is also a growing legion of Intellergy villains who live and work undercover side by side with us as acquaintances, coworkers, supervisors, and even self-proclaimed allies, teammates, or spouses~ The densities of these beings resemble that of stars in gravitational collapse relentlessly drawing matter and energy into them until they become the equivalent of a Neutron Star or a Black Hole from which not even light can escape~ Even though the majority of them are completely unaware themselves of their nature, self-absorbed in their dark Intellergy, too many of these entities are intentive masters at the manipulation of language and imagery attachments that are at the very core of the Human Disease~

Witness the woman who is repeatedly beaten and abused by her husband, only to return again and again for more. Each time he makes her believe he is sorry and promises it will never happen again~ In the workplace every day, workers succumb to abuses of power in the forms of degrading verbal abuse, grueling work schedules, or impossible production standards from superiors with virtually sociopathic tendencies~

With apologies for this foray into these shady, obscure regions, Blindsight reminds us that there is no lasting power or evolution here~ Dark Intellergy continually feeds on itself like the Ouroboros until it is consumed and immersed into _____~ The best defense is the simple, fearless recognition and observation of these entities, along with the relentless intention of universal love and actualized awareness~

Simple stitches in the fabric of a mind
Unfold magnificent tapestries
In the halls of a Multiverse~
— MWUSO

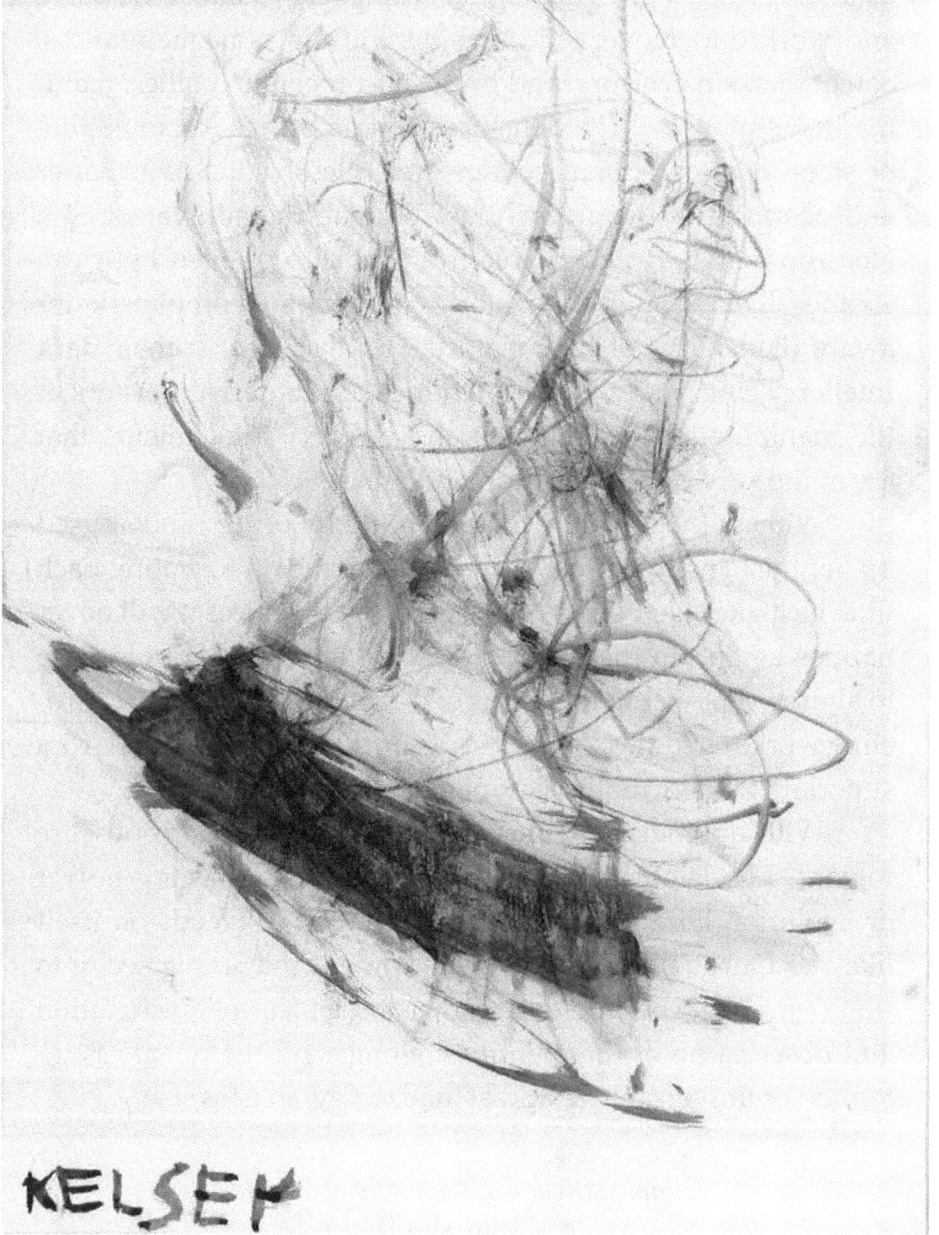

Long de Tiene by Kelci Auliya, age 3 years

XXVI

A Light Matter

The only difference between the saint and the sinner is
that every saint has a past and every sinner has a future~
— Oscar Wilde

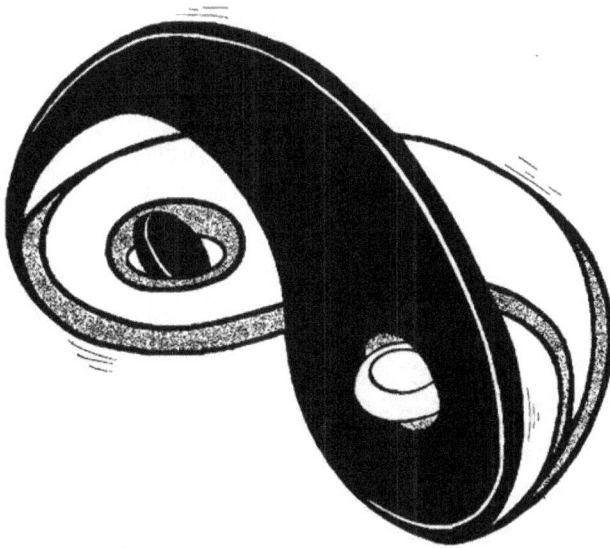

Sithainn Yin Yang by Kelci Auliya

Following such a shadowy Thojourn, it feels necessary to seek an Intellergetic balance~ Oscar Wilde's quotation implies that within every saint is a sinner and within every sinner is the potential for redemption~ The Chinese have a symbol called the Yin and Yang used to represent the phenomenon of apparently conflicting and opposing forces as complimentary and interconnected; ultimately, interdependent in the Multiverse~ We speak often of duality in these chapters~ Yin Yang implies something that has been called *dualistic-monism,* allowing that all things exist in a dynamic system composed of both aspects, and they tend to be identified by the dominant forces they express~ Ultimately, one cannot exist without the other in the One Mind; one cannot have shadow without light~

We wish to remind the reader that the darkest of Intellergies originally sprang from the intention of _____ through the Cosmic Inflation, and these aspects reside in all thought, imagery, objects, and actions in One Mind~ At the same time, Illuminated Beings not only currently exist in the Terra Signature, but their numbers are literally exploding at an exponential rate in the millennium~ Our humble voice arising now at the time of the Great Shift is but one of thousands approaching millions; these are beings called the Light Workers or Light Bearers~

The most expanded, actualized, and Intellergetic beings inhabit human bodies on Earth, but their similarity to common humanity effectively ends there~ Most of these entities exist in seclusion on mountaintops, in caves, along great rivers, in monasteries, in Native American or Aboriginal villages and other remote regions of solitude~ These locations are proactively and consciously chosen for their separation from the distracting densities of the Qen4 aspects of the Terra Signature and the Human Disease~ Practically all of their time is spent in the meditative state of Samadhi, Satori, or Nirvana, immersed in _____~ These beings spend their manifestations infusing the intentions of love, play, compassion, and peace Intellergetically into One Mind and the Terra Signature~ These practices maintain and enhance the harmonic balance between the Light Forces of expanded awareness and the Dark Forces of contracting densities~

There is a word in Indian religions, *dharma,* that has no real translation in Western languages, but in many traditions it has come to mean one's destiny or purpose in this manifestation~ Some versions give the meaning *to hold or to sustain~* The dharma of these highly evolved entities is the result of Intellergetic choices at their point of transition to continue to exist in denser physical forms, facilitating the transitions of Energy Signatures on evolutionary paths that can benefit from these intentions~ There is one belief system in Mahayana Buddhism that believes Buddha consciously left his body not to reside in heaven, but rather to remain in spirit form on an earthly level, assisting and simplifying evolution and awareness until every last soul had attained enlightenment~ Buddha had consciously chosen to be the last being to enter Nirvana~

On your path of expansion and enlightenment, journey with supreme confidence that higher Intellergy is always present to draw from and enrich your travels~ Remember, you travel only until such time as you realize that you have been fully immersed within your destination the entire time~

In this period of the Great Shift, there is, as we have stated, an exponential growth in Light Bearers within the Terra Signature~ These beings represent a very broad spectrum of Intellergy~ There are the worshippers, supplicants, and disciples found in the churches, synagogues, mosques and monasteries~ At this harmonic, they are found in the presence of priests, pastors, rabbis, imams, gurus, and other practitioners of the meditative arts~ The great majority of these Energy Signatures have had no truly realized awareness or contact with expanded Qen5 intelligence as of yet, but many of them, like our earlier friend Horace, possess a knowingness that there is *something greater than themselves*~ Others have actually experienced brief enlightened awarenesses that have been rapidly collapsed into the attracted density wave functions of pre-existing belief systems~ These fragile Intellergies in the grips of their Human Disease become easy prey for Intellergetic criminals posing as spiritual or self-help saviors~

Bhagwan Shree Rajneesh was a mystic, guru, and spiritual leader in the latter twentieth century who amassed a significant following during his life that has continued after his death~ His *Dynamic Meditation* techniques captured the imaginations of many, including the extremely wealthy with its elements of love, play, humor, awareness, and creativity~ His book, *From Sex to Superconsciouness*, gained him the nickname of the *Sex Guru* as he advocated a lifestyle of celebration rather than asceticism~ Rajneesh encouraged his followers to forsake their earthly attachments for spiritual reward by giving all their possessions to him~ Before his death, he had collected ninety-three Rolls Royce automobiles with a self-proclaimed goal to have 365, one for each day of the year~ His ashrams and communes have been linked to everything from immigration fraud and tax evasion to drug abuse, assault, and sexual aggression~ Addicted to nitrous oxide and valium, he dictated his last three books under their influence~

Such was the articulate nature of Rajneesh' teachings, with their harmonic resonances of great and wonderful truths, along with his beatific countenance and intense charisma, that countless Energy Signatures have followed his teachings~ Far too many of these found themselves in material and Intellergetic ruin~

Also, in the twentieth century, Christians witnessed the rise of Jimmy Lee Swaggart, Jim and Tammy Baker, Jim Jones, and David Koresh, all of whom amassed wealth and influence with their self-proclaimed gifts of prophecy and union with God~

We once had a very close Christian friend who repeatedly invited us to their Sunday morning service to see and hear their amazing minister, who had "brought so many lost souls to the Lord." We had continually and respectively declined her invitation, but finally agreed to go on one condition: we would go and listen to her minister if she would promise, for just this one service, to watch and observe the individuals in the congregation during the service~ What we witnessed was a handsome, smartly dressed man with swept-back dark hair and just the right amount of grey, with a well-worn King James Bible flopping open in one hand~ The other hand extended a long well-manicured finger and a hand with a not inexpensive ring~ He spoke boldly and heartfelt while gazing off into some internally visualized *beyond*~ He was simply the classic southern evangelist, and the collection plate rewarded his zeal and passion~ We could sense Intellergetic centers in our own Energy Signature resonating harmonically in terms of densities accumulated in a youth exposed to Southern Methodist and Baptist cultures~ At this point in our evolution, these resonances had now become only points of observation rather than attachment, and they also became a fortuitous opportunity for further dissipation of those pre-existing densities leading to more clarity and expansion~

After the service, we asked our friend what her impressions of the congregation had been~ She remarked that they had almost, to a one, appeared to her as lost souls with faces filled with anxieties and suppressed fears, looking to this pastor as their shepherd to lead them to safety, love, and salvation~ At the same time, she saw how this man she had come to respect was taking advantage of these people for his own per-

sonal gains in money and influence~ My friend never returned
to this church~

~~~*~~~

*There is an old Zen parable that begins with a Brahman
high priest in the holy city of Haridwar on the banks of the
Ganges River in India~ In his early beginnings, he had
been a humble student, spending his days in meditation
and studies mastering Hindu philosophy and history~
However, in his early manhood, he became afflicted with
the attachments and densities of the Human Disease and
began using his gifts to chart a course for wealth and power
within the derivative religious establishment~*

*In time, he rose to the highest institutional levels, living
in a magnificent temple surrounded by many servants~ He
wore robes of silk with brocade of gold~ He ate the finest
foods and drank the rarest wines~ Daily he stood before
audiences of supplicants expounding with eloquence his
teachings of Hindu practice and was rewarded in kind
with adoration and gifts~*

*One day he overheard some of his priests talking of
an old Buddhist monk who lived by a great river deep in
the forest~ They spoke with reverence and enthusiasm of
tales of the monk's amazing healings, his visions, and his
miraculous works~ The high priest felt he must know this
man so that he could share with him his vast understanding
and the possibilities of wealth and power available to one
such as he~ And so he sent his fastest runner into the forest
to find and return with this legendary man of the Buddha~*

When the old monk arrived at the great temple, he was offered a bath, clean robes, and a wondrous feast, but he respectfully declined~ He did accept a simple platter of dates and grapes with a cup of mint water~ When the high priest met the monk, he immediately began attempting to impress him with his deep and extensive knowledge of Hindu philosophy as he gestured grandly around his fabulous temple~

The old Zen monk sat peacefully and attentively in Lotus on the floor of the great hall, with the slightest of smiles in the corners of his eyes~ At length, the Brahman exhausted his expansive knowledge and presentation, and then he asked the monk for his thoughts~

The monk rose slowly and lithely from his Lotus and bent to remove his worn, vine-woven sandals~ He placed them carefully on the top of his wispy-haired head and began to laugh heartily~ He then turned and danced joyfully out of the temple all the way back into the forest~ He was never seen or heard from again~

~~~*~~~

In character, in manner, in style, in all things, the supreme excellence is simplicity~
— Henry Wadsworth Longfellow

Blindsight would warn to beware the *Guru Trap* from both sides of its manifestation~ The above examples present the aspects viewed from an observer position; however, do not underestimate the Intellergetic temptations, densities, and attachments that appear in transition into Qen5 harmonics~

The concept of the Guru Trap came to us through one of the highest Intellergetic beings in our energy sphere~ As a young man, he had walked across Afghanistan and India seeking enlightenment, and after finding and winning the confidence of a realized master, he spent two years in a cave meditating~ For one of those years, he never spoke and only used a chalk tablet for communication~ He related that each day as he returned to his physical consciousness from his meditations, there were up to seven or eight people sitting around him who had heard of the *holy man* who lived in the cave~ They had come to ask him to teach them; to become their guru master~ The enticements were extraordinary to take on these supplicants, who would tend to his every need and fill his every wish, but his answer was always the same, "Go and meditate; you are your only guru, so be that." This being lives in the world but not tied to its densities, and as of this writing he is an entrepreneur, inventor and author~ In more than thirty years of knowing him, we have never heard him utter a single negative thought~

As the reader strives to ascend through the turbulences of Qen4 Intellergies on the path to expanded space, one will encounter partially awakened Energy Signatures that are still ensnared by the Human Disease~ They can be most readily identified by their overuse of I~Me~My~Mine language and a forceful intention emanating from them that insists on acceptance of their belief system and recognition of them as superior entities~

These examples and myriads of others down through history portray human Energy Signatures' futile efforts to define, manipulate, and exploit dualistic relationships with _____~ Whether through book, edifice, institution, doctrine, or persona, all attempts other than simple immersion inevitably fall into spiraling, entropic cycles of dysfunction and decay~ We observe the valiant struggles of Islam, Judaism, and Christianity for relevance in a millennial context; however, their unabated attachment to and exploitation of dualism is more and more revealed for the illusion that it is~

None of this is to discount or deny the works and contributions from the manifestations of the fully expanded Signomes who have emerged and continue to emerge within the Terra Signature from the densities of these religious traditions~ It is their evolution and transition through to higher harmonics that resonate most with the Energy Signatures searching for more expanded and clearer truths, distancing themselves from the densities and tensions within their religious contexts~ They are Intellergy Forming arguably some of the most impactful Qensional bridges of the Great Shift~

Examples of the *pretenders* have been offered as cautionary Intellergy for the readers crossing the Dimensional Bridge~ We have found that once just one of these Intellergy criminals is recognized and observed, it proves very difficult for other such manifestations to create any kind of density in an Energy Signature that is intent on awareness of the true nature and immersion into _____~

Just below the harmonics of the highest of the Light Bearers exists a broad strata of resonant, Qen5-influenced beings who are Qen4 active in the Terra Signature~ These entities are known by their works and they will never advertise, market, or solicit a fee for their efforts~ Some of these are full time at their endeavors, contributing as mentors, healers, authors, artists, philanthropists, environmentalists, and others~ Light Bearers at this vibrational level are in constant resonance with Qen5 Intellergy, incapable of succumbing to the tensions and densities of the millennium or the infections of the Human Disease~ Historically, we have seen the likes of Buddha, Mahatma Gandhi, John C. Lilly, Mother Teresa, Franklin Merrell-Wolff, and the Dalai Lama who manage(d) to move through the world without being attached to or overwhelmed by it~ Much like the scarab beetle worshipped by ancient Egyptians for its ability to crawl through the dung without being soiled, these Light Bearers bring the powerful evolution of Intellergetic intention and awareness into their every breath, while gathering very little or no density~

Also appearing in the Terra Signature as we pass through the Great Shift is a another, much larger contingency of Light Bearers with intimate connections to higher Intellergetic awareness~ Most of these beings are millennial Indigos and, in many cases, exhibit extraordinary sensitivities manifesting as psychic awareness, quantum healing abilities, and even minor telekinetic tendencies~ The central differentiating quality of these entities is that they still possess large accumulations of Qen4 density and symptoms of the Human Disease~ These densities, along with their sensitivities, have been a part of these Energy Signatures ever since their manifestation in the Multiverse shortly after the Big Bang and the Cosmic Inflation~ Theirs is a constant struggle to understand and evolve through the inner conflicts, tensions, and densities within the harmonics of their Intellergy~ Driven by their intimate connection with expanded consciousness, they fight relentlessly for their true natures to emerge~

Because these Light Bearers have such high concentrations of expansion as well as contraction Intellergy, they possess the greatest potential to act as bridging examples for the transition to Qen5 consciousness~ They are exceptional communicators; they are very simply the quintessential Qensional Bridges~ As we approach the Quantum Point, these beings are making use of the global internet brain to network, organize, and blog~ They are building confluences of evolutionary Intellergetic Beings with an intentive goal to raise the harmonic resonance of the Terra Signature, wipe out criminal Intellergy Forming, and create a new paradigm for creativity, awareness, entanglement, play, and love~ The reader would be well served to search out these portals, find those that share harmonics and resonances with one's current evolutionary frequency and look for ways to perhaps join and contribute Intellergetically~ If for no other reason than observation, this effort can be of significant value for one in search of contact with expanded awareness~

From its inceptions, Blindsight's intention has been to stimulate and facilitate the expansion of realized consciousness in the Terra Signature~ It is these Bearers of Light tirelessly working to raise the Intellergetic frequencies of creative intention who are driving the Great Shift toward an entangled understanding of our true natures in a Multiverse immersed in

_____ ~

Grian Saoghal by Erin Nicole

XXVII

Nomes

A Genome is fundamental genetic material of an organism that makes up DNA, leading to determination of basic traits such as sex, color, size, and so on~ It will not be necessary to go into any further detail regarding genomes, as the term is easily researched if the reader so decides~

For the purposes of Blindsight, the word *genome* is simply an inspiration for some of the terms conjured for the purposes of Thojourns in this writing~ As we mentioned early on in the simulation, we want to not only begin to use words with a little more elegance, but also encourage others to explore the use of words of their own creation to expand the language~ Most of these have already been used previously; however, we would like to spend a little more time expanding their understanding~

Signome

The self-actualized consciousness of a Signome is inde-scribable in Qen4 language and beyond human understanding; however, we will make an attempt~

A glimpse of a suggestion might be the state of a fully re-alized spiritual master in a state of fully-expanded space with no movement, no vibration, no motion, no name, no body, no identity, no object~ It is the state of Enlightenment, Nirvana, Samadhi, or Satori~

One defining element of a Signome is that this Qen5 state is still fundamentally entangled with a Qen4 form~ Signome is a metaphysical, transitional *platform* from which a fully entangled being may engage intention and choice to return to Qen4 manifestation, advance to Qen5 strata for further evolutionary challenges, or bypass all unnecessary evolution and remain in simple oneness with _____~

Terra Signature

Ancient ancestors of the Sapiens species celebrated innate knowledge and belief that the earth is a living, breathing, self-aware being consisting of intelligence and energy~ Just as the human form is made up of trillions of living cells, molecules, and microbes that work in concert to form the entity, so Gaea is also~

Heaven is under our feet as well as over our heads~
— Henry David Thoreau, *Walden*

As you walk upon the sacred earth, treat each step as a prayer~
— Black Elk

The nexus for connecting Qen4 Intellergy to Qen5 enlightenment is the pure, unavoidable blend of the human Energy Signature with the Terra Signature~ Simply from a physical standpoint, approximately 60 percent of the adult human body is water, while about 70 percent of the earth is made up of H_2O~ The resonance is undeniable~ The harmonics are immediate~

In the millennium, accelerated Intellergy generated by technology has overwhelmed that of the Terra Signature in sheer volume of information, innovation, and environmental impact~ Albeit, the resonance of earthly Intellergy is at a lower frequency from a perception standpoint, but it contains a richness of deeper harmonies that can facilitate the transition for Qen4 beings through synergy in concert with the inherent Gaian intelligence~ Great shaman adepts, such as Black Elk, had cultivated the ability to act as conduits for this transmission of earthly intelligence they related to as their source~

As Energy Signatures, we all are intimately in tune and in touch with the earth at all times, even if we are not always aware~ Without our intention or choice, we must participate in gravity, the atmosphere that rejuvenates us, the food sources that sustain us, and temperature ranges that allow us to survive~ Through our Intellergy Forming abilities, conscious or not, we are inextricably connected with the Terra Signature~ Blindsight would offer that heightened intention and attention directed at this amazing synergetic opportunity will significantly enhance our prospects for the highest energy and intelligence to reach critical mass as the Quantum Point of the Shift approaches~

At the dawn of self-aware intelligence, Energy Signatures knew they were inseparable from the Terra Signature they inhabited~ When a Lakota Sioux chief in the nineteenth century was shown a treaty marking a map of territory to be ceded to the U.S. Government in exchange for horses, blankets, and trinkets, he found it absurd~ His response was, "That would be like asking fleas to divide up a dog." These indigenous and aboriginal tribes knew and acted from the intrinsic awareness that every part of their existence was inseparable from their environment~ Spiritual belief and mysticism sprang from keen awarenesses and observance of the myriad forces and energies around them~ It is simple to understand from this perspective how these Intellergies could find insight, strength, understanding, and inspiration from the *spirit* of the sun, an eagle, a river, or a crystal without compromising the inherent knowledge that behind all these forces and energies remains one Great Spirit~ These practices go back millennia, before Moses stood at the burning bush~

The compelling quantum leap now upon us in the form of the Great Shift is fueled by the accelerating density of millions of human Energy Signatures seeded with Indigos, Light Bearers as Intellergetic Bridges, and a burgeoning growth of Signomes~ All of this Intellergy is linked by the global brain of the Internet~ There are *Pods* of these beings beginning to link, join, and pool their Intellergy to facilitate a critical mass and accelerate the Shift in the face of great turbulence~ This turmoil and confusion was foretold and expected~ Both Energy and Terra Signatures thrive on flow and attraction~ Accumulated densities resist evolution and change~ Once a consistent vibratory state is reached, its tendency is toward entropy, not a higher energy~ Great energy and intention is required to manifest more expanded exceptional realms~ Intellergetic Beings must face and overcome tremendous resistance and turmoil generated by these extremely dense energy structures to reach, establish, and maintain them~ The Human Disease is filled with these densities in the form of belief systems such as cultural, political, sociological, institutional, personal, or spiritual~

The Terra Signature is literally fighting for its very survival, and it is drawing from the Intellergy of the Millennial Beings as they draw upon it~ There is absolutely no certainty for the outcome of this Quantum Point~ The Great Shift and the Multiverse have *no dog in this hunt*~ Either resulting outcome ultimately becomes simple fodder for observation~ The One Mind is simply the unbiased observer until called upon with Intention~

Will self-aware, conscious, loving, playful Intention be the driving force behind this quantum leap in manifestation, or will reactive, entropic turbulence overcome?

Blindsight would challenge you to choose your intention sooner than later~

Be the change you wish to see in the world~
— Gandhi

Terranome

Terranome is another unknowable, indescribable entity that nevertheless engages Thojourns vital to the expansion of Intellergy~ It implies a fully entangled, realized, and expanded oneness with _____ in a global state of Satori~ Terranome is the one true goal of the Light Workers in hopes of reaching a global concentration of Qen5-actualized Intellergy generating a Quantum Point shift to expanded space~

Please place the vision of Terranome at a destination point on your Thojourn *map* and apply your best intention along this at once playful and turbulent path~

Solanome

G. I. Gurdjieff often spoke of the sun as the most evolved intelligence in this part of the galaxy, because it is its Intellergy in the form of heat, radiation, gravity, and light that fosters, sustains, and evolves life and creates the opportunity for self-aware consciousness~ Sun worshippers have often been viewed as primitive and simple-minded~ Perhaps some of these had a more expanded perception than was outwardly apparent~

Solanome, like Terranome, is unknowable and is put forth as another Thojourn for Intellergy expansion~ An Energy Signature on its pathway through to space may find a kindred resonance in Terranome or Solanome~ Such kinship will be a knowingness on the path to _____~

Siansadh Astar by Kelci Auliya

XXVIII

Thojourns

I've been in my mind
It's such a fine line
That keeps me searching for a heart of gold~
— Neil Young

Smooth seas never made a skilled mariner~
— Fortune Cookie

Let's have some fun with that~
— Bruce Springsteen

As stated earlier in the Quantafiers, all things perceived arise from thought~

Some of the most famous *Mind Voyagers* include Nicolaus Copernicus, Galileo Galilei, and Albert Einstein~

Copernicus was able to journey through his mind to put forth his heliocentric idea that the earth revolved around the sun, in the face of prevailing geocentric acceptance that the earth was the center of the universe~ He was able to realize this without the benefit of a telescope, mathematical instrument, or any other measuring device~ Even though his theories placed the sun at the center of the universe, they still flew in the face of those who were convinced that role fell upon the earth~ During his life, he faced ridicule, imprisonment, and possible death at the hands of those who were threatened by his thought experiments~

Galileo Galilei spent a large part of his life imprisoned because he refused to refute Copernicus' ideas~ However, in his time he fell into many reveries of thought to explore his theories, as well as his art~ Perhaps the most well-known is his vision of being in the windowless cabin of a smoothly sailing ship~ He imagined the act of tossing a ball with a mate or observing the motions of a fish swimming in a tank in the cabin~ He imagined that he would toss the ball the same way whether the ship was traveling or anchored~ The implications for physics and understanding of motion were paradigm-shaping thoughts~

Albert Einstein, at about the age of sixteen, conjured up a thought experiment in his mind, imagining what it would be like to race alongside a beam of light~ From this journey of thought spawned the later confirmations of gravity, motion, the space-time continuum, the curvature of space, $E=mc^2$, and countless extrapolations that define many of our perceptions and technologies today~

In 1960, Robert Heinlein wrote *Time for the Stars* about Tom and Pat Bartlett, identical twins taking part in Earth's first interstellar space mission~ This thought journey inspired two other twin brothers, James and Gregory Benford, to not only become physicists, but also science fiction writers~ In 2013, they hosted the Starship Century Symposium, advancing prospects of human expeditions to the stars~ James has postulated a high-powered solar *sailship* propelled at up to 10 percent the speed of light by microwaves~ The thought experiment born of a novel in the 1960s is now one of the best prevailing models to carry humans to other galaxies~

Please know that in no way do we place ourselves in the company of these giants~ Thought explorers from Heinlein to Asimov, from Tolkien to Rowling and countless other science fiction/fantasy authors have voyaged across their imaginations sparking possibilities, many of which have found their way into brilliant innovations and technologies~ We simply express our gratitude for their pioneering thought explorations into unbounded concepts, inspiring the rest of us to attempt the same~

Although our Thojourns are on a heading for a different destination, nevertheless with these voyagers' tacit blessing, we embark~ One reminder we would offer is to please remember that each thought along the path is a journey in and of itself, to be savored and not hurried~ After all, the journey is as important as the destination~

We would now offer some Thojourns that have served us well~ It is our hope they may do the same for you, or spark a creative response for others to follow~

Breathe deeply, smile in your heart, and remain present in the now~

Enjoy~

The Mind of a Hurricane

To set the stage for this first Thojourn, we would like the reader to take a few moments to contemplate the energy and intelligence necessary to form, evolve, and maintain a giant storm such as a hurricane, a typhoon, a cyclone, or a monsoon~ For most of us, the Intellergy aspects are simpler to imagine, since many times these meteorological events are referred to in the media as *storm systems*~ On the other hand, humans continually retain a certain arrogance and superiority with respect to intelligence, insisting that if something was actually intelligent, it should be able to communicate in such a way that a human could understand~ If that hurricane is so smart, why doesn't it just speak English or Mandarin or Farsi, for gosh sakes?

From a Blindsight and harmonic perspective, human intelligence is an extremely low vibrational reflection of Qen4 with its densities, dualities, tensions, and perceptions of matter~ Belief systems and preconceptional thought streams cycle through the Energy Signature at such low density resonances that a sort of *shield array* is formed around the energy sphere, preventing other informational frequencies from harmonizing with the Intellergy~ A piece of music played, for example, in the key of C would not be able to blend with scales in the key of C#~ With continued cycling of density and attraction in upon itself, an Energy Signature becomes a virtually closed Ouroboros circuit~ These entities reach a point in their manifestations where very little if any new Intellergy can be absorbed~

An Intellergetic Being on an awakening path has the potential to openly attract and embrace One Mind in all of its manifestations of intelligence and energy~ As alluded to earlier in the discussions of Terranome and Solanome, the Multiverse is One Mind entangled in all of creation~

~~~*~~~

When we were a child about twelve years of age, Hurricane Carla blew through our small Texas town~ We can remember looking out and watching as trees were buffeted until they broke~ The corrugated, sheet metal roof blew off our neighbor's garage and powerlines lay crackling in the street~ We could feel the immense power of the storm as the gales roared around us, rattling boarded-up windows and doors~

All of a sudden, we were in the eye of the storm and we can remember our father taking our grandmother home, believing that the worst was most likely over~ In the calm, we walked outside to stand in absolute stillness, where only minutes before chaos had reigned~ There was knowledge that all around us, violent turbulence continued unabated~ There in the epicenter of the cyclone was the peaceful heart of the storm~ Within less than an hour, the storm was raging once again and we knew, even at the age of twelve, that we had sensed an Intellergetic presence~ In a youthful state of wonder, we had stood at the center of something much greater than our self~

~~~*~~~

The breeze sings in the trees~
The rain dances on the roof~
The mountain stands as a sentinel~
The sun smiles on the land~
None of these things were known
Until we observed them~
— MWUSO

~~~*~~~

Observe a satellite radar image of a great storm and take notice of the amazing organization and intention of purpose present~ Such a tempest represents the Intellergy of the Terra Signature far beyond human capacity to understand or control~ One can find the same aspects in the body of a stone, the moon, or a constellation~ There is expansion of Intellergetic space in every observable object, event, image, thought, and action~

Now, please enjoy *becoming* the storm~

A Perfect Storm

Please begin with the Breathe~Play~Now exercise for a few minutes~

Imagine an epic storm, cataclysmic in its nature, irresistible in force, virtually infinite in size~

Remember, proceed slowly~
Breathe and let the path unfold; after all, an epic storm takes time to build~
Within this storm, I am a raindrop~
I am condensed from a cloud of potential and probability, where I existed only as mist, a cloud, a mere possibility; a fog of unlimited potential~
Now, I am drawn, falling relentlessly toward a gravitational center, a metaphorical earth, if you will~
I am blown and battered, tumbled and tossed~
My shape is constantly changing and evolving~
I am liquid~
I am frozen~
I am snow~
I am hail~
I am sleet~
I am melting~
Once again, I am thrown into the fray~
I collide with other raindrops~
We meld, forming larger drops, only to break apart again and again~
The turbulence is relentless~
I am afraid~
I am lost~
I am angry~
I want this to stop~

Every effort is futile~
I feel like I am dying~
Suddenly, I fall into the ocean~
There is no more upheaval~
There is no turbulence~
I am at peace~
I am endless~
I am ocean~
Here, I remain in the deep~
Completely in harmony~

_____~

Over time, the storm passes and the sun emerges, beginning to evaporate moisture back into the atmosphere~ There is no resemblance to the particular drop that merged with the ocean, only molecules of hydrogen and oxygen~ High in the reaches of the atmosphere, these billions upon billions of molecules begin to join and collect, forming new clouds~ The cycle of collection and absorption increases until the cloud reaches sufficient density and electrical charge to stimulate the next thunderstorm~

The raindrop is reborn~
The cycle continues~

~~~*~~~

In this little journey of thought, the storm is a metaphor, representing the densities in an Energy Signature accumulated by all the contacts, traumas, sicknesses, successes, failures, and such that have risen or *evaporated* from the endless, peaceful source of the *ocean* of consciousness and energy that is

_____ and gathered in the *storm clouds* of its field~ These are illusory and fleeting~ All storms must pass~

In this Thojourn, breathe, play and refuse attachment to the *storm*~ All the thunder, the lightning, the wind, and the rain pass away~ The storm moves on, leaving nothing but calm, clear skies~

I am sky~

I am ocean~

_____~

The Lost Chord

Before we begin our next Thojourn, we would like to explore one of our favorite quantum principles, String Theory~ String theory is essentially a concept relative to gravity in a quantum sense~ Instead of looking at the universe as a particle zoo, this theory's proponents prefer to look at the universe as one-dimensional strings, and as strings, they vibrate~ The vibrational states still have apparent mass, charge, motion, and other properties ascribed to particles~

String theory is actually championed by many as one of the Holy Grails of Quantum Mechanics known as the Theory of Everything~ Since the birth of the field, massive physicist mind/brains have been convinced there is one unifying model that can describe all forms of matter and all of the fundamental forces~ Having spent most of our manifestation as a musician, String Theory holds a special place for us, and it is what led to our next Thojourn~ However, some of the other qualities of the theory stimulate many aspects of the Blindsight view with its inclusion of an infinite potential for other universes and its love of the paradox~

> *Music in the soul can be heard by the universe~*
> — Laozi

~~~*~~~

When we were a mere ten days old, our mother returned to her position as bookkeeper at the local bank in our small Texas town~ We were placed in the care of Veo, an African-American woman, and her family, where we spent most of our days and many of our evenings~ Veo was one of six lead singers in a thirty-voice Southern Baptist choir, and on Wednesday evenings we went with her to services held in an old barn that sat out in a cotton field on the edge of town~ She always put us in a seat behind the pianist who, in our memory, must have weighed at least four hundred pounds, but when this woman played, we felt like the earth moved and the voices became stratospheric~

This was the birth of music in us~ During the day, Veo would find rhythm in the vibrations of the washing machine, and she would dance with us as an infant around the house to the beat of a spin cycle~ Later, when we were older, we joined in the washer dance~ When vacuuming, she could find harmony in the whine of the machine, singing gospel and blues chops around the changing hum~ Outside in the yard, we were amazed as Veo carried on whistling conversations with the mockingbirds or sang while her old bulldog, Buck, howled along in harmony~ We learned instinctively from our earliest manifestation that music can be found literally everywhere if one is willing to listen~

<div align="center">~~~*~~~</div>

The musicality of String Theory for us brings an intimate elegance to the concept of a Multiverse infused Intellergetically with melody, harmony, dissonance, rhythm, tempo, modulation, dynamics, and dance~

Pythagoras came to the conclusion that music was the ordering of the world, and he is traditionally given the credit for discovering music's harmonic structure~ In the exact numerical relationships between the harmonies, he saw intelligence within reality~

. . . rhythm and harmony find their way into
the inward places of the soul, on which they
mightily fasten, imparting grace and making
the soul of him who is rightly educated graceful~
— Plato

String Theory's vibrational characteristics dovetail nicely into ancient Hindu metaphysical tradition~ In Nada Yoga, the nada represent sound vibrations that make up the universe, rather than particles or matter~ In practice, music and sound are applied in the role of intermediaries and transition energies for immersion into higher consciousness~

The sound *OM* (also referred to as AUM) for the ancient Hindus was a sound considered the *mystical syllable* or the *cosmic sound~* Other translations of the sound include: the essence of life, the cause of the universe, the infinite, and the truth~ Many of the ancient yogis claimed it was the sound uttered by Brahman at creation~ Spoken correctly out loud or internally in meditation, it is said to contain all of the open sounds of human capability~ It is also believed to stimulate the creative powers of the universe~

Once again, the earliest voyagers in thought describe with compelling clarity some of the most exciting theories on the quantum fringe, with no access to particle accelerators or complex mathematics~ So take heart, my Intellergetic friend, and know you have the same connection to Multiversal immersion as these great masters~ It is our position that the path is clearer and less cluttered, for it's lacking the complexity of science and math~

Here's hoping you find a wonderful transition melody to play in our next Thojourn~

Domhan Giotar by Kelci Auliya

A Multiversal Guitar

For our purposes, let us take a Martin D-35 guitar built in the mid-twentieth century~ The hardwoods available at that time had not been compromised by over commercialization or deforestation~ As a result, the tones coming from the resonance and response of these woods brought to life in the hands of a master luthier rival even the best Stradivarius~

In this Thojourn, our D-35 is suspended in space without attachment to any stand or apparatus~

The guitar is at rest, capable of unlimited possibilities of expression and response~ It exists only as XPotential~ It is a Multiverse~ It is a probability cloud~

According to quantum principles, a Multiverse refers to an infinite number of universes, including our own~ These are also referred to as parallel or alternate universes ~

_____ plucks the low E string on the guitar with full intention, stimulating the infinite XPotential of energy and information that has been lying dormant up to this point~

The inspiration followed by the conscious, self-aware intention and the pluck of the note is a Big Bang~

The ensuing vibration is Cosmic Inflation~

A Multiverse is born~

The E string begins to vibrate~ If one were to observe this string closely in slow motion, one would find vast numbers of sympathetic vibrations undulating along the string, each with its own energy and information infused with the intelligent intention of _____~ Within these undulations are other smaller vibratory states, harmonic, resonating universes unto themselves~ The E string alone is its own Multiverse, but it does not stop there~

Immediately, the entire body of the guitar begins to resonate in reverberating sympathetics to the initial pluck~ The exquisite hardwoods enhance the tones filling the ambient spaces within the body, focusing out the aperture in the center~ Every cell and molecule in the wood are vibrating and Intellergized, adding depth and intelligence to the tones~

At the same time, each of the other five strings have begun vibrating sympathetically with their own internal undulations~ Within the waves are many levels of deeper resonances and harmonies, all of them creating other Multiverses~ Each of these is a harmonic different from the one created with the original pluck, but inextricably entangled with its pitch, intensity, and tone~

The atmosphere surrounding the guitar is now also responding with harmonics and vibrations~ Each resonance and harmonic are Big Bangs of their own, spawning ever more Multiverses~

~~~*~~~

At this point, we would like to expand this particular Thojourn and apply it with personal Intention for an individual Energy Signature~ Please note that it could be related to almost any perception of the Multiverse~

~~~*~~~

In this Thojourn, the Multiversal Guitar represents the Energy Signature of an individual human with each string embodying the most powerful I~Me~My~Mine density attractors of Intellergetic personality and identity~

The high E string embodies the Intellergetic center of the being, the Impersonal I that is the true nature of the Energy Signature~ These are the finest and purest of densities that have been attracted for transition to higher Qen5 resonance~ They are made up of the most fluid, malleable, and transitional Intellergy forming the bridge to Signome expanded consciousness~ These are the harmonics brought into manifestation through mindful Intention, creating unbounded joy and unconditional love ~

The B string represents the core I of this being~ This is the central identity most associated with the name of the Signature~ As such, it is most often perceived as the whole being not only by other observing Energy Signatures, but also by this Energy Signature itself~ Just like so many would observe and hear the whole guitar, instead of perceiving it as contributions of the collective strings and body, most will respond to the harmonics and resonances of this I as representative of the whole being~

The G string will symbolize the inspired, creative, and passionate I~ This personality aspect is the social, professional, spontaneous, and active face to its manifestation~ This I is the face that, for this being, is portrayed to the world~ The densities attracted to this personality system are the driving forces for success, career, family, adventure, and enjoyment for this being~

The D string characterizes the serious, committed, and dedicated traits of the Energy Signature~ This aspect manifests in expressions within religious, political, environmental, and social belief systems~ These are harmonics with resonances leading to action within Qen4 timbres~

The A string will denote the insecure, fearful I harmonics of the personality that has attracted densities resulting from pain, disappointment, trauma, and failure~ This identity system can react at any time from any perceived threat, confusion, misunderstanding, or physical pain~ When stimulated, its harmonics become so powerful that they are capable of manifestations of reactions ranging from depression and contraction to those of urgency and flares of temper~

The low E string will represent the lowest, strident, discordant, and dense frequencies of the Energy Signature~ These are the dark harmonics of the deepest fears and most traumatic experiences of the being~ When roused, the most violent, angry, and hateful manifestations may arise~

As with an actual guitar, each string responds to the intensity of the pluck~ A real guitar string would react with volume and longevity of the tone~ Likewise, our Multiversal Guitar harmonics would react with intensity and relative prolonged manifestation~

As with the real guitar, one string being played resonates with other strings on the instrument~ For example, an inspired pluck of creative epiphany Intellergy on the G string could be followed by a resonant harmonic on the high B string in the form of some sort of awakening to new and exciting potentials~ At this point, the Signature will experience higher frequency perceptions with proactive XPotential and mindful action to fully realize their vision~

~~~*~~~

The Thojourn of the Multiversal Guitar may be applied to virtually every human thought, word, or action~ Its application and growing understanding in concert with *Breathe~Play~Now* Intention can rapidly accelerate the conscious evolutionary process for an Energy Signature~

This Thojourn implies an Observer and, as a result, the Multiverses go on and on as each observation crushes the wave functions in its path~

Obviously, this Thojourn can be expanded endlessly~ However, if we imagine our own universe's Big Bang as _____ plucking a *quantum string* with fully realized intention, we can begin to envision the process of creation~ We can also see that virtually everything we perceive in our so-called Multiverse is merely harmonics and resonances from the initial *pluck* and the Cosmic Inflation~ Most of what we may attribute to *free thinking* or *free will* is actually reactive response to the consciousness, energy, and intentive forces set in motion at the singularity~

It is even more likely that this Multiverse was in itself a reactive result of a prior event in a preceding universe, as mentioned in the Big Bounce earlier~

After all, please make no mistake that the Xpotential for unfettered thought and free will are infused into this Multiverse of reactions~ This Intellergy awaits only simple, present and playful intention~

Play on~

Colormind

This Thojourn is best taken in a dark space and works well in conjunction with the *Zero Gravity Position* developed by NASA for healing and resting their astronauts~ Very simply, the way to imagine this position is to see yourself sitting in a recliner, reading or watching a movie~ Now just push that position all the way back to prone so that the head and shoulders are slightly raised with the knees a little higher than the heart~ This can be done easily with pillows~ Please make certain that that the pillow under the head also supports the shoulders in such a way to align the neck and facilitate an open breathing passage~ *A pillow under the head alone will force the head forward, stretching the neck and constricting breathing*~ For the legs, a longer pillow or cushion of higher density foam or padding is recommended, to support the entire length of the legs~ The height of the leg pillow is a personal choice relative to body type, but the best way to find the position is to feel when the spine completely relaxes, flattened with the surface~ This will be an almost immediate feeling of relief, so please trust this and adjust until you have reached an optimum position of physical relaxation~ The Zero Gravity feeling is not exaggerated if one does this correctly, and one will feel the effects of this position instantly~

This Thojourn works extremely well if one has experienced a turbulent, high-density day, and it can also be taught to children for facilitating bedtime or naps~

We will offer this next Thojourn in the first person as we experience ourselves~ We would propose that if one would like to join, please try recording the exercise also in the first person, so as to engender a more participatory meditative environment~ Hearing the narrative spoken in one's own voice and invoking the sound of *I* creates harmonics and resonances in the Intellergetic Mind/Brain that make for a natural progression of the creative visualization~

When played for a child, read in first person by a familiar, loving voice, the reading becomes more like a game the child is watching from a sideline~ The child then has the free will to join in when and if they choose~ This simple act of free, personal choice is very empowering for a child in the three-to-five-year age range who has very little personal power in their life~ It has been proven to dramatically increase self-esteem, confidence, and creativity~

We have also found that playing soft, instrumental music with a tempo just under that of resting heart rate is extremely conducive to maximize the Thojourn~ *Super Learning* research has found tempos between forty and sixty beats per minute with slower rhythms relax mind and body functions and facilitate alpha states in brain waves~ Classical Baroque music from the likes of Bach, Vivaldi, and Handel are proven sources~

Remember that all you have and are is NOW~ _____ is all around and within you~

Here now is the script: (Go slowly, very slowly~ Allow the breathing to unfold along with each of the visualizations~ Remember that this is a peaceful journey~)

~~~*~~~

I close my eyes~
I take very deep, slow breaths~
I am playful in my heart~
I place the slightest of smiles in my eyes and in my mind~
I breathe very slowly and deeply~

With my eyes still closed, I look up into the space in my head just above my eyes~

I see my favorite color~

I continue to breathe and be patient while the color appears~

I see only my color in my mind~

Now, I choose to brighten my color~

I make it as bright as I possibly can~

I am patient~

I am slow~

I breathe~

My color is brighter~

The glow is a living light that breathes with me~

The light is playful and loving~

I now make the color into a circle, an amazingly bright and beautiful disc~

I breathe and I play~

I am slow~

I am here NOW~

I make my circle into a ball~

The ball is amazingly beautiful~

I begin to send intention for it to spin~

After a while, I reverse the spin~

I speed it up~

I slow it down~

I am present~

I am free~

I am at play in the colors~

I now dissolve the ball and allow the color to expand, filling my mind~

I see my head filled with this brilliant, bright color that is solely my own~

I let the color begin to flow . . . . downward into my body~

My color spreads through my chest and arms, down into my legs~

My body is glowing with the intelligent energy of my color~

Along with the breathing and play, I feel profound joy and peace~

I am my color~

My color is spreading outside the boundary of my body . . . spreading out into the room . . . filling it with my incredibly bright, energetic, and intelligent light creation~

The color-light is pulsing and beautiful~

I am infinite~ I am entangled in my true nature~

My color expands, reaching out into the night, passing the moon and beyond the stars~

At the speed of my thoughts, my color has become the universe entangled with One Mind~ There is no body, no motion, no time~

I am space~

~~~ * ~~~

It is highly recommended that upon waking from this Thojourn that the traveler attempt to draw or paint what it is that they saw and experienced~ Understand that this is the very essence of the creative process: Imagine, Believe and Achieve~ When employed with a child, the shifts in creativity, self-esteem, and confidence can be remarkable~

These Thojourns can be endlessly created and adapted by imaginative and free Intellergy~ One can find many examples of meditations and creative visualizations available from many sources~ It is always recommended that one create the thought journeys using the Intellergy centers of one's own being~

The infinite nature of these voyages is just what makes them so compelling and powerful for one who aspires to escape the meat suit and embark to a presence that is fully entangled Intellergy~

Because _____ is completely entangled in the Multiverse, there is the ever-present opportunity to be _____~

Cease to be the harmonic and become the player~

It is through Thojourns and brain/mind observation and exercise that we approach oneness with _____~

We are the creators of our Multiverse~

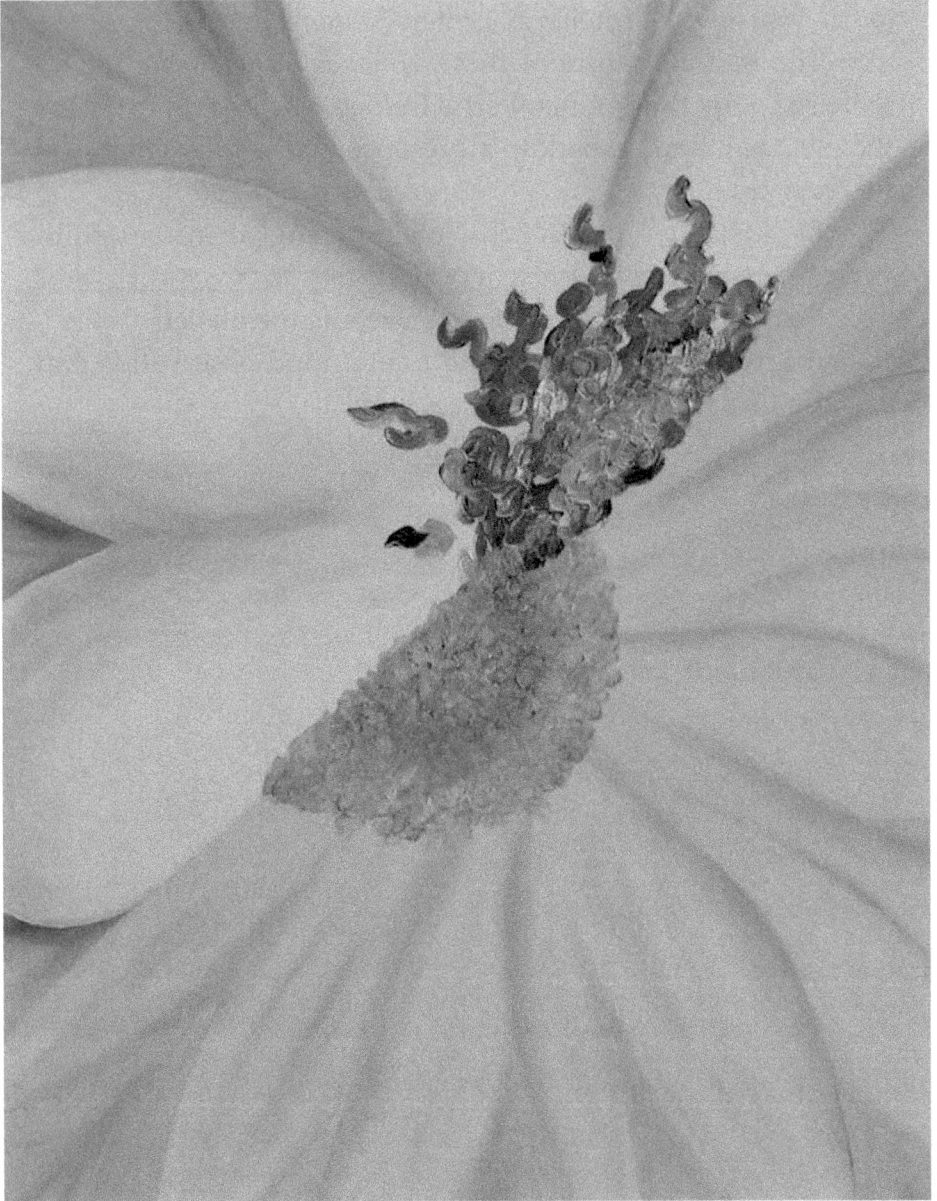

Loinn by Erin Nicole

XXIX

Intellergy Forming

Just after the Big Bang, something called the Cosmic Inflation, or the Inflationary Epoch, took place during which the Multiverse expanded at superluminal speeds, effectively establishing the fundamental beginnings for the growth of structure~ This period lasted only the Nano-blink of an eye from 10^{-36} seconds to about 10^{-32} seconds after the singularity~ The Multiverse continued and continues to expand at a much slower rate, perhaps influenced by the Higgs Field~

For Blindsight, what is significant about this astonishing point in cosmological history is that in that instant, the Multiverse was infused with a flow of infinite energy, consciousness, and the intention of _____~ This flow of Intellergy is the stream of endless creation~ The Multiverse is in a constant state of renewing itself within the One Mind, reacting to the intention of the singularity~

A present, self-aware Intellergetic Being has the singular ability to be a part of this flow of creation while not being *in* it~ Through the diligent and extremely difficult work of eradicating the densities of the Human Disease and engaging Qen5 intention, XPotential creativity becomes reality~ This ultimate level of creativity, we will call *Intellergy Forming~*

As the Multiverse constantly renews itself, it is renovating your Energy Signature without your permission or participation, and this progression will continue until the inevitable dissolution of your manifestation's Intellergy, with its re-absorption into the multiversal pool~ The challenge and the question becomes whether one wishes to be an Energy Signature that is a robotic product of dense, reactive intention or an XPotential Intellergetic Being?

> *What happens is a continual surrender of himself as he*
> *is at the moment to something that is more valuable~*
> *The progress of an artist is a continual self-sacrifice, a*
> *continual extinction of personality~*
> — T. S. Eliot

> *I saw the angel in the marble,*
> *and carved until I set him free~*
> — Michelangelo

Intellergy Forming is adventure~ It is entertainment~ It is experiment~ It is the singular most fun an Energy Signature can have~ When density has dissipated to a certain extent, one begins to experience reality as more fluid and malleable, instead of fixed and solid~ Objects, events, beings, and actions take on aspects of energy and intelligence that begin to respond as present, proactive intention is applied~ First, awareness of this phenomenon will appear with practice in Thojourns or meditations where one can begin to *improvise* with higher mental states, unencumbered by Qen4 densities and duality~ The Colormind Thojourn makes a great platform for this type of expansion~

> *Imagine for a moment the Intellergy Forming in the*
> *minds of J. R. R. Tolkien creating Middle Earth or J. K.*
> *Rowling's Wizarding and Muggle Worlds~*

There are arguably an infinite number of portals for connection and immersion into One Mind and further into _____, because One Mind and _____ are omnipresent, ever-awaiting for our instantaneous and conscious entanglement~ We are always fully entangled, we just rarely are aware of it~ It is helpful to remember that we are constantly involved in the process of creation~ Since the Big Bang and the Inflationary Epoch, the Multiverse has been unfolding, infused with the creative intention of _____ endlessly reinventing, revealing, and observing itself~

Our manifestations are the result of that creation, but we are continuously being re-created~ From a physical body standpoint, red blood cells are fully replaced every four months, liver cells every five~ Your lungs are only six weeks old, your skin is about two to four weeks old and your heart about twenty years old~ These are only Qen4 aspects, but they are reflective of the creative process of which we are all a part in this ever-changing, ever-expanding universe~

The universe is an accelerating Intellergy event of cumulative intelligent energy coalescing into informational densities that Qen4 Energy Signatures perceive as mass, motion, and thought~

> *Thoughts become things~*
> — Mike Dooley

We have hopefully learned from the Buddha that attachment and desire are the great factors of the Human Disease leading to duality and tension~ However, as Energy Signatures we are inextricably a part of the Cosmic Inflation's creative flow~ Through rigorous practice and intention, we have the unique opportunity to step into and influence this flow becoming creators ourselves~ For many, this is a striking beginning on a path to the awakening of unity with _____~

So God created man in his own image, in the image of
God created he them; male and female created he them~
— Genesis 1:27

Almost all creativity involves purposeful play~
— Abraham Maslow

Music, dance, theater, painting, drawing, writing, and sculpture make up the great general categories of art~ Of these, live performances of music, dance, and theater come closest to bridging Qen4 performance with Qen5 Intellergy, mostly because of their spontaneity elements~ A truly inspired concert, enactment, or choreographic interpretation can move others to higher, more expanded states of being~ Because these singular live acts of creation are performed in the *NOW*, they most reflect the creativity of the Universal Mind, since they have different energy at each performance~ For example, a performance of Beethoven's Fifth Symphony in London's Royal Albert Hall will be, in many ways, completely different than the same piece performed at the Teatro alla Scala in Milan~ The musicians would be completely different, as would be the construction of their instruments~ Tempo, volume, and nuance would be personalized at the same time, along with the acoustics of the venues~ At the same time, the audiences listening to the performances would be demographically different in terms of culture, age, taste, and so forth~ From an Intellergetic perspective, these would be two entirely different works of art~ If one were to record these concerts, they would be effectively collapsing the wave function of their probability clouds~ One might read a review of the recording that it had *captured the magic of the performance~* The mere fact that it was somehow *captured* validates the collapse~ Although such recordings are incapable of evolving, they have often been known to inspire further and even higher states of creativity~

Paintings, drawings, writings, recordings, and sculptures are fixed densities; however, their stimulus has often sprung from the highest inspirational consciousness and energy~ As such, these pieces are most capable of generating very moving experiences of a Qen5 nature for those viewing, reading, or listening to them, as each Observer filters the Intellergy through their own density spheres~

> *The position of the artist is humble~ He is essentially a channel~*
> — Piet Mondrian

> *Making your unknown known is the important thing~*
> — Georgia O'Keefe

Not everyone is capable of creating works of art, great or otherwise~ There is the brilliant mechanic who always wished he could play guitar; however, an epiphany on the design of the configuration of an engine might revolutionize an industry~ The software designer who had always dreamed she might have the gift of becoming a great painter might create the next application that would facilitate greater doctor-patient connectivity~

The reality is that each Energy Signature is constantly fashioning an amazing, ever-evolving work of the creative masterpiece that is one's life in the form of Intellergy~ Densities, collapsed waveforms, attractions, and repulsions are constantly collected and transformed in endless progressions of creation~ This is the _____ imperative infused into the Universal Mind at the Big Bang and the subsequent Cosmic Inflation~ Consciously or unconsciously, you are creating every moment of your existence~ Sadly for many, 100 percent of the creative choices made to fashion their present reality are reactions to the infection of the Human Disease, engendering the illusions of free will and creativity~ We literally spend almost all of our manifestation as blind slaves to Qen4 turbulence, duality, and misconception~

Once again, it is our hope that the reader is at once shocked and challenged by this concept~ The Blindsight vision offers up remarkable pathways to freeing one's Energy Signature from the bonds of this robotic existence~ Opportunities exist to embark on roads of creating each and every moment of your reality~

> *You're in the unknown. And from the unknown,*
> *all things are created~*
> *You are in the Quantum Field~*
> — Joe Dispenza

> *You are creating your life right this moment*
> *by how you think~*
> — Ramtha

> *We are what we think~ All that we are arises with our*
> *thoughts~ With our thoughts we make the world~*
> — Buddha

> *As a man thinketh in his heart, so is he~*
> — Proverbs 23:7

Self-aware consciousness with self-actualized intention is our one and only instrument for composing the symphony that is our existence, the only brush for painting the masterpiece of our *NOW*, the only body tuned to the ultimate dance of our creation~

Creation is~

Creation flows~

You may have heard the reference *stream of consciousness*~ It is into this Qen5 current that we are able to step if only we openly, willingly, and consciously choose~ Staying present in the flow of this stream is the most difficult part of the journey, as the perils of Qen4 instability and attractive energy concentrations constantly threaten to overturn the fragile craft of heightened Intellergy~

Once one has come into contact with higher intelligent energy fields, most will find themselves continually kicked out by the sheer overwhelming flood of expanded awareness~ At first, there is simply no frame of reference for processing this expansion~ There is no Qen4 language to describe or categorize~ There are no mathematics to formulate or conceptualize~ There are no objects or imagery capable of anchoring any kind of position, frame of reference, or motion~

If the first and subsequent immersion experiences are true, most will be consumed to the point of addiction, with the desire to return to the more expanded states, along with a yearning to extend the presence as long as possible~ What is required is the repeated and diligent hard work of the mind/brain~ Make no mistake that it will be the hardest endeavor one will ever undertake; however, after the first immersion, it will be perfectly crystal clear that there is simply no greater reward for success~

Desire is our anchor to Qen4, but it is also the vehicle for transition~ It is the desire for expansion, awareness, true nature, and oneness with _____ that fuels intention~

At the Qen5 level, there is primordial, fully realized, self-aware intelligent energy with which we are already entangled that simply awaits our intention~ It is this Intellergy that is our instrument for creation of our present *NOW*, which is the basis for the next *NOW* and the next~

Begin to cherish and celebrate that you are Creation; you are the Creator~

You are not separate from some faraway, unknowable supreme being~ You *ARE* the supreme being inasmuch as a red blood cell is you~ You are fully entangled already, just like the cell~

Be that~

Know that~

Trust that~

Like the cell, your manifestation will pass~ The Intellergy will transition and transform, never lost, only reappearing in the endless metamorphic cycle of creation within the One Mind~

There should be no fear, although there seems to be just that in the beginning immersions~ Consider for one moment that there is only one thing that is not changing at near-light speed in every nanosecond of your existence~

That would be your name~

Please consider that your name is a cleverly designed collapse of your Energy Signature wave function given to you by another collapsed wave function at the time of your Qen4 inception~ Densities of formulaic, derivative, and predictable attractions instantly begin to coalesce around this small, seemingly insignificant Intellergy~ The insidious I~Me~My~Mine~ contamination of the Human Disease is implanted~

As mentioned earlier, the most terrifying experience of our own manifestation came as a result of our first foray into Qen5 space, as we completely lost all sense of body and self~

Knowing our true nature as one of fully expanded, self-aware space and energy, we became intimately fearful we had died~ Fortunately, we were able to maintain self-awareness long enough to pass through this layer of turbulence and have the most amazing vision of our infinite XPotential~ It has been this revelation and the many that have followed that have fueled our evolution~ It proved to us the attachment to the *identity* or *personality* that is our name became the anchor of the densities in our Energy Signature~ If there were no name defining our swirling, fluid energetic *body,* it follows there would seemingly be less to be attached~

One can vastly accelerate one's evolution and the transition to Qen5 Intellergy by beginning at once to distance oneself from attachment to one's name~ Inevitably, in a restaurant, coffee shop, or some other public environment, someone will ask for a name to be called when your table or beverage is ready~ One of our favorite games is to simply and playfully make up a new name each time~ In different situations, we are currently known as Bosco, Ozzie, Shuggie, and our personal favorite, Scooey~ We attempt to conjure an entirely different persona on every occasion~ While introducing an entirely new dynamic to the *Searching for Signs of Life* game, it has also proven to be an excellent tool for dissipating name attachment densities and Intellergy for some serious fun~

Attempt at this moment to intimately understand that when you act as an infinite Intellergetic Being, undefined and unrestricted by name definition in an uncollapsed probability cloud, the XPotential for full entanglement is endless~

> *You create your life through the inner power of your being, whose source is within you, yet beyond the selves that you know~ Use those creative abilities with*

*understanding abandon~ Honor your selves and move
through the godliness of your being~*
— Jane Roberts

The best way to predict your future is to create it~
— Abraham Lincoln

*If you create your life with love, your dream becomes a
masterpiece of art~*
— Miguel Angel Ruiz

Know no fear~ Fear is a powerless illusion~
Create *NOW* boldly with abandon~
Love unconditionally~

NOTES

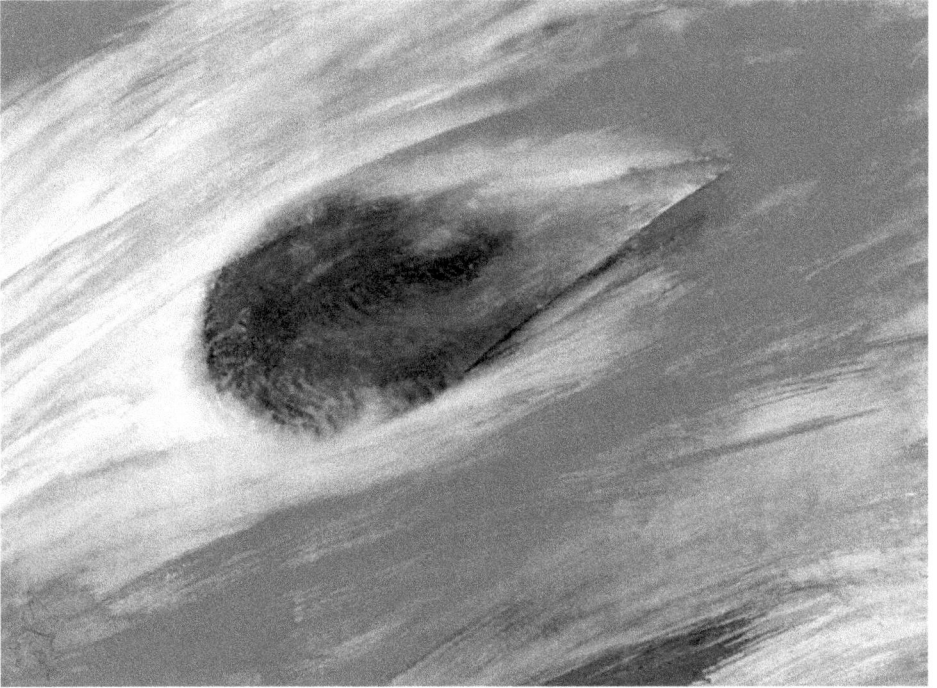

Foirfe Gailleann by Erin Nicole

XXX

A Blindsight Life

What you are is a force — a force that makes it possible for
your body to live; a force that makes it possible for your
whole mind to dream — you are life~
— Don Miguel Ruiz

This is actually our third attempt at creating this simulation over the last ten years~ We have had working titles such as *Quantosophy* that had a certain pretentiousness we could not endure~ We also tried *Quantelligence* and *Quantergy*, which fell short of our intentions~ For an instant, we even thought of using "Quansciousness," but for some reason it always made us think of candy~ Try our new *Quanscious Bar; It's Quanschy!* You can probably imagine why it did not make the cut~

We realize at this point that this chapter is the central reason why each attempt found lacking had nothing to do with a choice of title~ We apparently required some sort of destination or culmination point~ On some level, we must have needed a journey's end to justify the journey~ Now we can see the journey was completely unneeded, as we have always been exactly where we should be~

The river is everywhere at once, at the source and at the
mouth, at the waterfall, at the ferry, at the rapids, in the

sea, in the mountains, everywhere at once, and that there
is only the present time for it, not the shadow of the past,
not the shadow of the future~
— Hermann Hesse

A journey, however it has been, and it is far from over~ The difference now is that we travel less and less with the perceptions of ourselves in a physical form and more and more as an Intellergetic being~ We have even arrived at the point where when we catch ourselves acting too *human* it is almost comical~

For example, recently we had hitched the trailer to our truck, and we were driving to the landscape yard to pick up a few yards of planting mix for our fall garden~ We had gotten about ten miles down the road when we realized that we had left our wallet with all our money at home; a very *human* thing to do~ We felt that immediate tightening in the abdomen, the explosion of frustration and possible options in the brain, along with the expletive of the moment~ Suddenly, as if a silent alarm had sounded, our higher awareness awakened, brought us into the present moment, and made us intimately aware that we were bringing all this tense human, reactive Intellergy into the *NOW* that simply had no place~ An immediate smile appeared in the Energy Signature with the realization that all our solutions lay before us~

For Quantum's sake, these solutions represented probability clouds of infinite Xpotential~ Naturally, the only option was to return to the ranch for the wallet, but now we were able to celebrate the journey in Motion Meditation with calm joyful presence, where in the past, it could have been a trip filled with anxious densities of tension and the regrets of our mistake~ The event served as another lesson in Intellergetic living that, in its simplicity, has made us less likely to overreact in similar situations~ In the spirit of Human Disease prevention, it is our goal to be in such a state of presence and actualized awareness that no reactive, formulaic, or derivative thought or action will ever arise~

One may counter: What about an emergency or an attack?

In the practice of Zen, there is a *Zen moment* when time feels as though it has slowed to almost a stop~ This is commonly reported in different ways by great athletes, race car drivers, martial artists, battlefield soldiers, and others~ For One Mind, every *NOW* is a probability cloud of infinite Intellergetic Quantum Points awaiting intention~ An Energy Signature that is present and in resonance with this Intellergy is capable of meta-human performance responses at XPotential levels~

As one evolves Intellergetically through higher frequencies, these Qentinuum expansions become more frequent and more easily anticipated~ In an emergency or an attack example, there are almost always Intellergetic flows in process well before any actual encounter that a present, aware Intellergy will sense and begin to temper their strategies from a proactive, intentional perspective~ Even in an instantaneous crisis, a calm, present Intellergy offers not only less chance of error or misjudgment from overreaction, but also limitless opportunities to create Xpotential solutions from *NOW*~

Along the way we have been amazingly fortunate, some would say blessed, to have come into and remain in contact with a number of highly evolved sensients~ These beings inhabiting human physical forms work quietly and mundanely in the Qen4 Multiverse with present and actualized Qen5 level luminosity~ Their works, creations, and intentions are continually altering the very fabric of Terra Signature reality preparing avenues, portals, and bridges for the transition of the Great Shift, and yet they remain virtually unseen and unaccredited~ For them, the act of becoming visible or creating a profile would not only radically diminish their power, but it would also draw to them forces and densities complicating their missions~

~~~*~~~

One starry night with a single candle and the smell of burning sage on the porch of our little ranch house, we were quietly sitting with one of these beings~ We were simply present, breathing deeply and playfully when we clearly saw all around us, above and below us, thousands upon thousands of Energy Signatures hovering around us in our awareness~ At the time, we remarked that we felt as though we were in an energetic arena~ There was an intimate happiness and welcoming energy, along with the sensation that we were suddenly somehow lighter in weight~ At no time was there a sense these were external meta-beings or oversouls~ There was no duality~ All of these Energy Signatures were very simply harmonics and resonances of our own true nature~ Our higher, One Mind Intellergy had been silently and patiently awaiting our simple acknowledgement to become a conscious, participatory part of our Energy Signature~

From that night on, we have begun to experience, move, act, speak, and feel more as an expanded presence rather than a physical body~ It has been remarkable that we now have begun to more easily observe ourselves acting like a human and it seems to always bring a smile, realizing this transformation is far, far from over~

As fortune would have it, we were contacted by another of our friendly sensitives a few days later and told them of our experience in the arena~ They calmly and confidently offered that we had been in the Cathedral of Souls~ This, for them and their kind, is obviously a common reality, and it was a tremendous boost for us in the confidence of our experience~

~~~*~~~

Intellergetic evolution is progressive, and it begins where we are as an Energy Signature when we become aware of the possibility of living from a Qen5 perspective~ It is extremely long and hard work~ For most, it is labor intensive for the entire manifestation and beyond, but the rewards far outweigh the sacrifices~ Most of us find there is a familiarity that comes with initial exposure to the higher harmonic resonances~ It manifests with a quiet understanding as though the experience has happened before; a déjà vu~ There is also an intimate feeling of joy and a peaceful reconnection with one's true nature~ After so much time spent exploring, experimenting, and observing our Intellergy at work, we have begun to notice that certain accumulations of density and preconception are more easily identified~ In the cases of the pastor in our friend's church or the forgotten wallet, we were able to observe the events in the *NOW* as stimulations of derivative density centers~

By simply applying some playful intention to the sources of these erratic harmonics, we were able to completely eradicate the dissonant Intellergy~ As we progress in observing the manifestations we have created, it becomes much simpler to spot robotic habits, unfounded fears, minor irritating situations, belief systems, and preconceptions~ These indications of the Human Disease are the focal points for evolutionary intention and transition to more expanded awareness with regards to these symptoms' impact on our Intellergy~ Merely shining the light of truth on these densities that completely enslave us and lock us away from infinite connection is enough to inspire us to do whatever is necessary to continue freeing ourselves~

Earlier in the book, we asked some questions:

Why do we have to pay our bills?

Why do we have to pee?

At this juncture, these are questions that should no longer require answers, only observation~ It has become increasingly joyful to pee or eat or pay a bill or walk or... As we send our intentions into a creation that is our manifestation, our perception comes from a sense that we are an Energy Signature *system* of sorts, and for some reason at this time, we experience ourselves as a *spherical* Intellergetic form in motion through the Qentinuum~ There is an inescapable sense of our universal entanglement~ The simple act of driving down a street or walking through a store conjures the Intellergy that we are the street; we are the store~

We are all things~

We are _____~

Please do not infer that we ourselves are always in this state; however, these Intellergetic states have accelerated in frequency and duration~ More and more often we are able to conjure these experiences with the confidence they will return again and again~ Our presence and energy have increased exponentially with an almost constant feeling of simple joy~ The now-familiar company of our true nature has formed an ever-expanding clarity~ With the simplest acts, such as tying our shoes or writing these lines, we now experience these intentions channeled through this higher Intellergy~ Our manifestation is a creation infused with a self-actualized, Qen5 confluence ever-entangled with One Mind and _____~

Reviewing these past few paragraphs, once again we feel that language has proven absurdly inadequate for expression of these higher states, and it is our hope the reader can appreciate our open honesty in this feeble attempt~

Throughout our journey of more than half a century living and evolving vicariously through the words and experiences of the masters, we have always known this as our destiny~ Looking back, it has been more of an expedition, but it has afforded the abilities and the opportunity allowing us this communication~

As with all expeditions, most of them result in failure, and it is those willing to learn from these shortcomings and rise again with renewed strength and commitment who succeed in the end~ The struggles against the relentless Human Disease, along with the turbulence encountered Intellergetically with any and all efforts to raise frequency harmonics toward Qen5 level resonance, have resulted in far too many Energy Signatures falling back into illusory Qen4 densities~ Someone once said, "When you feel like quitting, remember why you started." You can always call on those astonishing Eureka and Epiphany moments to fuel your next attempt~

> *Try again~ Fail again~ Fail better~*
> — Samuel Beckett

With homage to ol' Sam Beckett, we can promise next time, we will fail better~

You are intelligent energy immersed in _____~

Be that~

Live that~

There is, very simply, no other way~

Grian Eun by Kelci Auliya

XXXI

A Blindsight World

In the spirit of the foregoing chapter, as well as the discussions on the Great Shift and the approaching Quantum Point, we would like to present the possibility of a Terra Signature elevated to higher Intellergetic awareness by the simple confluence of a significant number of Intellergies in critical mass bridged with Qen5 conscious intention~ Using the great meditation experiment in Washington, DC, described earlier as an example, we would propose the opportunity for generating a Quantum Leap in terms of global awareness and intentive action~ In physics, a quantum leap is defined as *a sudden large increase or advance; the sudden jump of one electron, atom, etc., from one energy level to another~*

Blindsight compels that there is a harmonic, resonant frequency associated with Intellergetic Xpotential states that alter so-called physical realities and perceptions~ These shifts may occur in an individual Energy Signature in the form of an unexpected change in one's attitude or perspective~ They could manifest at even a cellular level as an unexplained remission event in a cancer patient~ Given the preponderance of evidence and examples we have presented here, isn't it worth the setting aside of derivative, formulaic, and preconceived belief systems long enough to diligently and intentively explore the possibilities of living in a world of beings intimately aware of their entangled, true Intellergetic nature~ Imagine an earth inhabited by beings who are present and aware of the illusory nature of One Mind's duality characteristics and its Human Disease~ This would imply a critical mass of global, collective consciousness dedicated to the eradication of the disease and the conscious, proactive evolution of human Intellergy~

This is the work of the coalescing pods of Light Bearers in the Global Signature, racing to achieve a worldwide Intellergy consciousness frequency to fundamentally alter the Intellergetic makeup of the Terra Signature at the Quantum Point~

Blindsight's sole purpose in the creation of this simulation is to support, contribute to, and further this effort~ In our opinion, there is simply no more important mission~

It may sound to some as overdramatic, something from a Sci-Fi film or a video game if we speak of the rise of dark forces, criminal Intellergy Forming, and the Human Disease; however, in the context of the Blindsight undertaking, there is very simply no questioning the reality of these threats~

~~~*~~~

There was a time along a great river that was the very lifeline of all who lived along and within proximity of it~ The river was more than a water supply; it was transportation and food supply for people, business, goods, and services~ An entire region for hundreds of miles and tens of thousands of people depended on the river for their very lives~

One of the uses of the river was for the transport of giant trees that had been cut and trimmed far upstream that were being sent to milling operations further south~

One day, there was a log jam that clogged the entire river and before word could be sent to the loggers upstream, literally miles and miles of huge logs had accumulated in the stream, cutting off all passage~ Days went by and then weeks turned into months~ Engineers and technical advisors strategized and experimented with every imaginable solution to no avail~ Crews worked day and night with all sorts of machinery and explosives in efforts to break the blockage loose, but nothing worked~

Surrounding economies began to suffer dramatically from lack of supplies and commerce~ Many who had lived for generations on the river were forced to leave~

Then one day, a lone engineer, weary and tired of all the tension and confusion, struck off into the hills above the river and found himself in the midst of a growth of huge pine trees~ On a hunch, he began to climb one of the largest trees and later, as he sat swaying over the canopy of the forest, he was able to look down upon the hopelessly clogged river~ In a moment of simple clarity, the engineer saw where the source of the blockage had built up~ As fast as he could, he made his way down and found a single stick of dynamite~ After clearing everyone away, he ran out onto the log jam and placed the single dynamite stick in the place he had seen from above~ With one relatively small blast, the entire blockage was freed and began to flow unencumbered down the river~

~~~*~~~

The accumulation of densities, the dualistic illusory nature of Qen4 perception, along with the Human Disease, have created an almost insurmountable *log jam* in the Terra Signature's Intellergetic flow~ Intellergy Forming criminals generate profit and power from the perpetuation of these illusions in complete denial of the ultimate futile, entropic, and destructive efforts in which they engage~

A collective Intellergetic union of enough expanded and entangled Energy Signatures can strip the obstructing forces of all power with the simple illumination of the illusion that is their density~ Just like the log jam freed by one small burst, Intellergetic Beings can begin to experience the uninterrupted flow of Multiversal creation with one simple immersion into

_____~

In the end, the mission is so very simple: recognize, know, and implement our true nature~

> *And you shall know the truth,*
> *and the truth shall set you free~*
> — John 8:32

Blindsight cannot and will not allow these Intellergies to drive the outcome of the Quantum Point into darkness without the most committed of intention for an illuminated XPotentiality; a Quantum Leap in expanded, entangled Intellergy~

This is the vital imperative Blindsight would offer, with all the urgency one can muster~ With it comes a great challenge and a responsibility~ Each Energy Signature must find and develop their own harmonics for this cause~ They must seek resonance with other Intellergies of similar densities to construct their own bridges to higher and more expanded awareness~ All Intellergy bridges have a common destiny, and the Quantum Point will reveal they are all the same fully entangled connection~

Simple stitches in the fabric of a mind
Unfold magnificent tapestries
In the halls of a Multiverse ~
— MWUSO

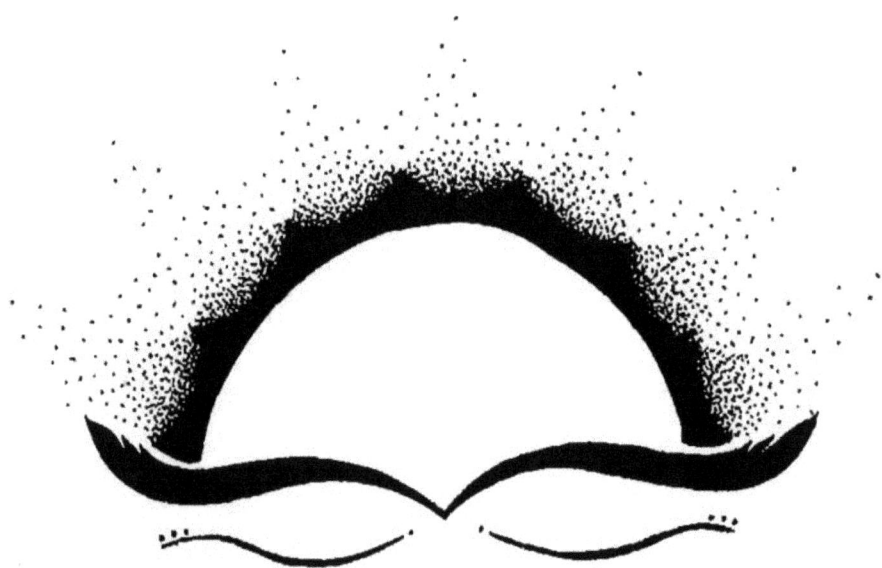

XXXII

To Believe
or Not to Believe

~~~*~~~

White Wolf stumbled and fell once again in the snow above the timberline on the frozen side of the Sacred Mountain, deep in the Bitterroots~ He could no longer remember how many times he had fallen at this point~ The young Nez Perce brave was three days into his Vision Quest~ After many nights chanting and communing with the elders, he had embarked on his climb~ Fasting, now praying, and now crying out to the spirits for the revelation that would reveal his true purpose for how he could best serve the People, he trudged on~

This morning's dawn came as a blaze of fire in his eyes, with a pounding in his head~ The ground beneath his frozen moccasins seemed to wave and flow like a molten liquid~ There came a ringing in his ears and through the din, he was beginning to hear the voices of the spirits on the mountain~ He felt light and powerful; he knew he was close~

As he looked up from his fall, he saw stretched out before him a large rattlesnake, almost frozen and very near death~

"Please help me," the snake pleaded with the boy, "I am so cold, and I feel I am dying."

White Wolf started toward the snake and then pulled up in fear of the scaly predator~

"Why have you stopped," the snake asked~ "I need only to be warm."

"How can I know you will not attack me?" the boy questioned~

"I am freezing and perishing~ How could I do you or any other harm?"

White Wolf, in his state of fatigue-induced illumination and bent on completing his Vision Quest, began to see and believe that the snake's peril was the opportunity for which he had been striving to commune with the great spirit and complete his purpose~

He bent and picked up the snake, wrapped him into his buckskin shirt, and began to make his way back down the mountain~ The rattlesnake was stiff and moved hardly at all~

Moving now into the trees below the snow, he began to feel warmer and he noticed the snake begin to stir~

White Wolf opened his shirt and began to remove and release the snake when all of the sudden the rattlesnake viciously raised up and struck the boy at his throat~

As the boy lay dying in the soft needles of the lodge pole pines, he spoke to the snake~ "Why have you killed me? I saved your life."

"You believed what you wished to believe," said the serpent, *"I was always a rattlesnake."*

— Adapted from a Native American legend

~~~*~~~

What one believes is arguably their most powerful stimulus for intention~ Beginning with the intimate certainty that one has a physical body standing on solid surface with total control of faculties, a platform is constructed for reactive Intellergy Forming that ripples out into every aspect of a manifestation~

The sky is not the limit
Your belief system is~
— Vanessa Feils

A belief system is nothing but a thought you've thought
over and over again~
— Wayne Dyer

You have heard of, and studied, various systems of
philosophy; but real philosophy is opposed to all systems~
— Frances Wright

A belief system is an Intellergetic *box* formed by an Energy Signature through repeated action, practice, and the attraction of sympathetic densities from other Intellergetic Beings in Qen4 reality~ Belief can be assigned to each and every density accumulated in the Signature~ These concentrations of intelligent energy foster illusions of structure, expectations of circumstance, and relationships of common resonance~

~~~*~~~

It was the fourth grade in Mrs. Pfeiffer's class during geography when she had pulled down a very large map of Russia for us to study~ One by one, she would call on a student and have them go to the front of the class to find a location on the map~

We looked on as our friends found Moscow, the Volga River, and St. Petersburg~ At last, it was our turn~ Mrs. Pfeiffer asked us to find Kamchatka~ We stood close, less than a foot from the map, searching with all our focus and intention for the city, the river, the mountain, something~ Suddenly, we noticed that behind us the entire class was laughing at us~

The teacher asked us to take a couple of steps back and look again~

There it was in letters twenty times the size of all the cities, rivers, and so forth~ **Kamchatka** *was emblazoned as a province in the northern far east of the country~*

We had allowed our belief, based on expectations from our friends' experiences with this exercise, to completely compromise our ability to apply the necessary intention to our search~

~~~*~~~

True believers become trapped in the cyclical nature of their beliefs, conjuring *leaps of faith* and justification for gaps in the performance of the system, never questioning the limitations being placed on their Intellergetic potential and possibility~ From the beginnings of this simulation, Blindsight has vowed to challenge every belief system and source of derivative, formulaic, and predictable thought~ These are the roots of the densities causing the Human Disease~

This is a chapter that has no need for expanded treatment of this subject; however, it has the value of refocusing attention and intention on the limits of belief~

Begin to observe and question your belief structures~ What is the harm if they are intimately true for you? Conversely, what is the reward for the discovery that a belief system is flawed or limiting?

A committed believer is ultimately enslaved by a belief system that is fixed, structured, and defined by institution, language, or doctrine~ The entrapment is enticing, as it affords a sort of security configuration for an Energy Signature storing fear-based densities with regards to randomness, unpredictability, and the potential of unfettered thought~

> *The true believer, no matter how rowdy and*
> *violent his acts, is basically an obedient and*
> *submissive person~*
> — Eric Hoffer

The Multiverse rewards all intentions of belief with reflective manifestation in Qen4 reality~ As John Lilly proposed, seek to transcend all limitations formed by belief~ Blindsight challenges each Intellergetic Being to be the present Observer, recognizing not only the density restrictions of your own beliefs, but also to remain vigilant and aware of density transmissions emanating from Qen4 Intellergies entangled in One Mind~

We are what we believe~

NOTES

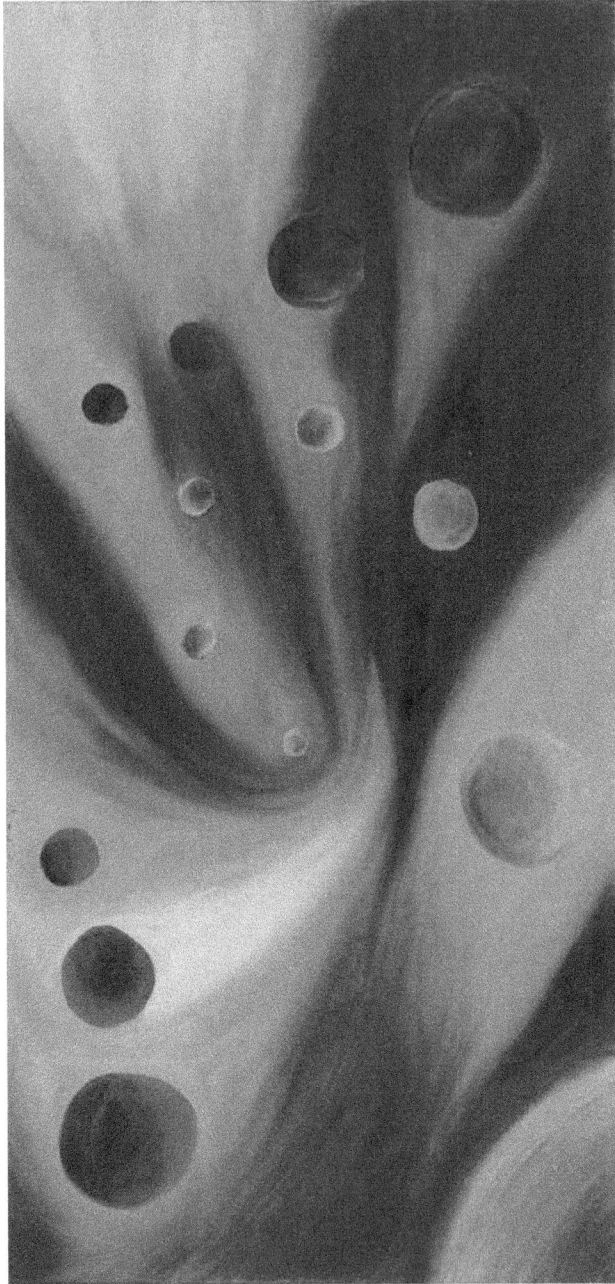

Miorbhailt by Erin Nicole

XXXIII

Transition

We are approaching the end of this simulation and for those of you who have taken the journey, we offer our deepest gratitude~

It is our hope you will pick up this book often and simply open it anywhere~ It is fashioned in such a way that after the first reading, it may take on a more random nature and current that may be *stepped into* at any place and time~ After all, it will always be *NOW*~

Perhaps there will even come a time when you may offer your well-worn, dog-eared copy to a random stranger in a park~

Throughout, we have offered suggestions for exploration, Thojourns, and research; however, at this point we would like to ask one last favor of the reader as we approach the close of our time together~ We ask that you read through the following Thojourn at least three times to familiarize yourself with the progression~

~~~*~~~

Please find a quiet place with some free time~ For the ultimate experience, find a remote hilltop at end of day and

sit facing west~ As you progress through the Thojourn,
feel *the earth beneath you rolling through space, spinning
around the sun, whirling around the galaxy, and hurtling
toward deep, fully expanded space~ In other words, be
where and when you are~*

Sit or lie down in a comfortable position~

Close your eyes~

Breathe~Play~Now for just a minute or two~

Now, say your name in your mind~

*Hear yourself say it clearly, and absorb the Intellergy of
this sound and know that this is the singular thing in your
entire life that has not changed or isn't constantly changing
at near-light speed at this very moment~ Understand this
is the signature of your Intellergetic Being at this point
in your manifestation~ It is the gravitational center that
continues to draw in accumulations of higher as well
as lower Intellergetic information in forms of densities,
tensions, harmonics, and resonances~*

Please say your name again in your mind~

Concentrate and see your body~

*Visualize your body fully clothed in comfortable
clothing in the same position you have chosen for this
Thojourn~ Give playful intention to this image that is
loving, calm, relaxed, and content~*

*Stay with this image for a short while~ Become
comfortable with it~*

*Now imagine your body in your present setting~ See
yourself relaxed and contented in this beautiful place~*

*Stay with this picture, breathing deeply, and playfully
becoming more and more calm and comfortable in your
surroundings~*

*At this time, simply send mindful intention that this
entire scenario is beginning to dissolve into an Intellergetic*

swirl of brilliantly glowing energy~ Your name, your body, the surrounding area itself are dissipating into more and more fluid, vague, and energetic shapes~

*Now you are seeing that all that remains is an Intellergetic field of Xpotential, a probability cloud of endless possibility~ At the edges of this field are **tendrils** of energy, like **flames** reaching into expanded space fully entangled with One Mind, ultimately immersed in _____~*

This your true nature~

~~~*~~~

Remain in this Thojourn as long as possible, absorbing the Intellergetic information of this frequency harmonic and resonance, storing this in your Energy Signature as a kind of *map* of a pathway for revisiting again and again~

This Thojourn is, with practice and consistent intention, such a simple transition into Qen5 perception~ All that is required is imagination and confidence in your truest form, which is fully expanded, entangled, intelligent energy~ Simply cast off any tendencies for attachment to physical forms of illusory mass with their attractions, tensions, and densities~

Embrace the simplicity~

In the end, there is no journey, no destination~ We are *NOW* and always what we are and always have been: an observational manifestation of One Mind~ _____ surrounds us, fills us, and permeates us~

There really is no *us* after all, right?

_____ is~

It has been, and still remains, our hope you have found in our little simulation stimulus and inspiration for your Energy Signature, some questions for which you will begin to crave answers and perhaps a glimpse into pathways through to your truest Intellergetic nature and its constant, present immersion into _____~ We also hope you will use these meanderings of an intermittent Signome as a reference to revisit from time to time for encouragement, focus, and refreshment~

We want to remind you that you are the creator of your reality and your universe~ *Choose your attitude* as the self-helpers might say~ Begin creating your day the moment you awake, with the deepest of breaths, the most playful of hearts, and the simplest realization of *NOW*~ Never forget that this creation of your day is realized in the next few seconds and the next~

In every situation in which you find yourself, you have the ability to become *NOW* and intercede, XPotentially altering the outcomes of your manifestations if you are mindful and present~ Use *Breathe~Play~Now* throughout your waking hours to bring yourself back to your present and infuse your moments with playful, focused Intellergy~

You have this amazing, infinite potential for creation of your own Multiverse~ You are capable of imparting infectious joy and energy into everything and everyone around you~

What is the worst that could possibly happen if you simply try?

> *It's better to aim for a star and hit a stump, than to aim*
> *for a stump and miss~*
> — Wild Bill Seals

If we were to find ourselves on our transition bed, would we want the energy we carry into our next manifestation to be burdened with residuals of fear, pain, anger, or confusion? We would rather hope to choose a perspective of wonder, adventure, and play~ At the moment of transition, as the I~Me~My~Mine~ perception of self abandons the lower frequency aspects of the Energy Signature, there is a conversion period of Intellergy that, for most evolved beings, can be experienced as presence in a new and different *place*~ Depending on the being's accumulations of densities and preconceived belief systems, these realms could appear as heaven or hell-like~ These experiences inevitably reflect the accumulated Intellergy of the Energy Signature with all its densities, belief systems, and preconceptions~ Mystics propose that this is a mechanism allowing the Intellergy to find a more familiar environment of transition reality on its path to immersion into One Mind~ Refer to Y. Evans-Wentz' excellent translation of *The Tibetan Book of the Dead* for some illuminating perspective on these concepts if you so desire~

We have the opportunity at every present moment to bring happiness, creativity, love, and play XPotentially into the next moment and the next~ Please remember, we are constantly cycling at near-light speed in and out of deep, expanded space, and each time we return with new, altered experience, energy, and consciousness~ We literally transition into a newly manifested Intellergetic being each time~

Live from your true Intellergetic nature and your beingness of _____~ This is a state of fully entangled, unconditional love~

Qen4 love is an emotion and, as such, it can become volatile or conditional in denser manifestations and lower frequencies~ It is too often a matter of opinion that varies from person to person or situation to situation~ How many times we have heard the expression, "I love you, *but*...?" This is not love if it is conditional~

Please meditate on the true, unencumbered, and infinite state of unconditional love~

Used with the intention of its creation, *Breathe~Play~Now* naturally engenders a state of unreserved love~ Consistent practice by an Energy Signature and spreading it to other Intellergies will literally raise the vibrational levels of the entire Terra Signature's harmonic spectrum~

Search out the pods and portals of the Light Bearers and the Millennials if the attraction is timely and right for you~ Join with them, if you are able, and find ways to impart positive, self-aware, creative, loving, and playful energy into every moment of your present manifestation and theirs~

Start where you are~ *HERE*~

Be when you are~ *NOW*~

Become the *OBSERVER* in your life~

Craft mindful thoughts that are fluid, clear, and creative~

Manifest your life with these~

Resist boldly every thought or action that is derivative, formulaic, or predictable~

Strive playfully to observe and understand every reaction~

Choose *REALITY* as your favorite toy~

Engage *INTENTION* as your most powerful tool~

Your *OBSERVER* is your very best friend~

Your *BREATH* is your beginning~

NOW is the greatest time in your life~

PLAY~

LOVE~

A wise friend once gave us some advice:

Keep a clear mind~

Have an open heart~

Seek to ease the suffering of others~

Speak only when necessary~

Speak to do no harm~

Speak only the truth~

***Truth is last, because truth is not always necessary,
and truth is also capable of doing harm***

ALWAYS REMEMBER~

You are unlimited energy in a state of complete quiet in a fine, fine blend, with infinite intelligent consciousness in a state of complete quiet~ There is no movement or activity; there is no arising of thought or object~

The next *NOW* moment of your manifestation will be created from some form of reactive or proactive intention infused into this *pool* of Intellergy~

What will your next *NOW* be?

These are your Transition Tools, your Power Tools for Blindsight Entanglement and immersion into _____~

Please begin, if you haven't already, to engage Intellergy Forming your manifestation with mindful, present, loving intention~

In that spirit, may we suggest you begin with SOWING LOVE IN YOUR PATH~

This Thojourn is seeding your flow with unconditional loving energy to be harvested by you and all who contact it~ You will find the most expanded space awareness in the acts of loving not just the beauty and truth on your trail, but being able to love yourself in the presence of unloved and unloving people and situations~

We have all seen, or even performed, the sometimes comical, always semi-serious act of someone spraying perfume or cologne into the air and then walking through the mist arms wide, eyes closed and allowing the gifts of the scent to surround us~

Imagine creating a field, medium, or Qentinuum of love Intellergy that your manifestation is moving through at all times harmonizing and absorbing the resonances of higher, loving frequencies~

The Multiverse is One Mind, One Being~ We are like blood cells in a body, grains of sand on a beach; we are stardust~

When we simply love ourselves, we love all things~

Love means that another's happiness is more important that your own~

So simple in concept; so difficult at times in application~

We are not suggesting that we all leap immediately to levels of the unconditional love of a Jesus, a Buddha, a Mother Theresa or a Dalai Lama~

The great message of Blindsight is that we KNOW we can do better~

We can be better as humans~

We can begin to cure the accumulated densities of preconception and belief that make up our Human Disease~

We can love better~

We can make this a better world~

Begin small and simple as you SOW MORE LOVE IN YOUR PATH~

Let someone go ahead of you at the line at the store~

Pay for the lunch of someone randomly behind you in the drive-through lane at the fast food restaurant~

Something as simple as focusing more loving, listening intention within a conversation with a friend, co-worker or love mate will bring such clarity and presence to the NOW of your existence and theirs~

Listen as the impartial Observer~ Try to keep all thoughts of judgement or how you will respond from your mind~ You will be amazed at how the person with whom you are speaking will expand their expression and reveal new depths of their being~ Trust that if any response is desired from you, it will be mindful and perfectly crafted in the spirit of speaking with the Transition Tools offered above~

Try to stay neutral in discussions and potential arguments~ It is so unnecessary to win any point of view in this dense Qen4 frequency~ Know that even in as little as an hour, it will most likely not matter, and the other person will have felt validated and respected~ You will have given of yourself, and the bonus is that your position is still intact!

Remember THERE IS NOTHING THAT IS UNLOVABLE when you are loving yourself~

Simply allow yourself to be in a state of GIVING as you move through your world~

Always GIVE first~

Instead of taking,

GIVE to life~

GIVE to others~

GIVE to yourself~

GIVE to your world~

GIVE to this moment NOW~

The Multiverse will return your generosity tenfold~

Learn to love your perceptions of your world and yourself~

The sign this is working for you is that you will feel all your senses heightened~

As cliché as it may sound at the moment, please trust that the world will be a little brighter, sound a little sweeter; in fact, you will find every moment you share with love is richer and more profound~

These kinds of Intellergy Forming actions naturally harmonize with love timbres and will dramatically raise your frequency resonances with the higher intelligence and energy of your being~ Such states create attractions of the rarest and finest of harmonies for the simple entanglement with the One Mind, and through this, immersion into _____~

The more you play with SOWING LOVE IN YOUR PATH, the more you find inspiration to raise your frequencies by tackling the more difficult tests of our love and patience, such as maybe a nosy neighbor, an irritating co-worker, or a pushy relative~

Other ways to accelerate resonance is to love yourself through difficult tasks or situations such as an aspect of your job you always dread, a trip to the dentist, or simply a menial chore like mopping or washing dishes~ These tasks become transformed from worrisome, tedious responsibilities to wonderful opportunities for evolution when undertaken with an intention of playful love infused mindfully into them~

We guarantee you that finding ways to SOW MORE LOVE IN YOUR PATH will literally change your life and launch your manifestation into the stratospheres of Intellergy vibration~

As we all strive to be a little better in our manifestation with every playful breath and every mindful, loving intention, we begin to experience our world unfolding in conscious evolution creating the highest of vibrational frequencies to feed the transitioning Great Shift~ The critical mass of uppermost resonance is our portal for evolution at the approaching Quantum Point~

There is true greatness within you~

There is great wealth, great love, and great experience awaiting the formation of your mindful Intellergy~

For most of us, greatness is an unfolding; it is a process, so for now . . .

Play better~
Love better~
Breathe deeper~
Be better~

As you depart this world of Blindsight, please do not forget to retrieve the belief systems and derivative thought streams you left at the entrance, if you so desire~ Examine them closely before you allow them to become fully reattached to your Energy Signature, as hopefully you may discover some alterations manifesting, as well as some lighter-density areas, more fluidity in flow, or clarity in perception~

In conclusion, we would leave you with our favorite song from an anonymous writer, as it expresses the entire core message of Blindsight~

Row, row, row your boat
Gently down the stream
Merrily, merrily, merrily, merrily
Life is but a dream~

~~~*~~~

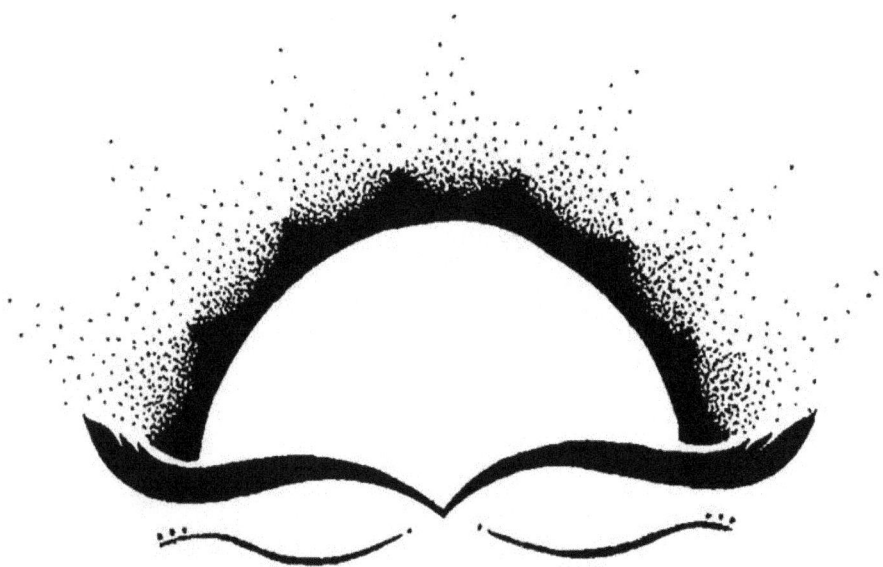

Expanded Glossary

1) _____

In Exodus 3:14, Moses asked God for his name and the response he received was, "I am that I am." Moses needed some kind of word to describe the indescribable~ "Jaab Sahib," the morning prayer of the Sikhs composed by Guru Gobind Singh, contains 950 different names for God~ In Blindsight, the use of a word or name for the so-called *Supreme Being* is, if not blasphemous, at the very least absurd~ For our purposes in this narrative, we will very simply use only _____~ If, for the time being, the reader would choose to insert some sort of personal identifying reference or word, they of course certainly may; however, it is our hope and goal that said reader will leave our discourse without the need for such naming of the indescribable; rather, we would engender the immersion into inexpressible beingness~ Any implication that we would offer up some sort of supreme being existing outside of our own oneness with _____ would be a complete misunderstanding~ _____ is the unknowable source of all manifestation and is infused within, throughout, and exclusive of all energy and consciousness~ _____ is the self-aware source of creation, observation, and intention that exists before, after, and permeates all Multiverses and dimensions~ Readers may choose to substitute "God," "Allah," "Brahman," "Atman" or _____~ The Blindsight position maintains that all these names imply a dualistic relationship between the self and a supreme being.

This will never be our intention~ Rather, we will attempt to establish that all duality (me~you, here~there, now~then, this~that, good~evil) is illusory~

The very endeavor we make to define or discuss _____ here is an exercise in absurd futility~ Nevertheless, oneness with _____ is the ultimate culmination of our journey~

2) Consciousness — For Blindsight, consciousness is intelligence in a state of self-awareness that one is cognizant of one's own inevitable and continual transition into other intelligent energy states~ Consciousness may have dualistic properties when focused on observing or manipulating an object or a process~ As in the case of energy, consciousness prior to the infusion of intention exists in a dormant state unto itself without an object in a quality of infinite Xpotentiality~

3) Density — Density in Blindsight terms refers to accumulations, obstructions, or concentrations of energy, information, and intelligence in the probability field of an Energy Signature or Multiverse~ These density fields form obstructions in the multiversal flow, as well as reservoirs for the gathering of compacted Intellergy~

4) Energy — The *Physics Hypertextbook* defines Energy as a scalar quantity that is abstract and cannot always be perceived; it is given meaning through calculation~ With respect to physics, it is defined as a property of objects, which can be transferred to other objects or converted into different forms~ According to the First Law of Thermodynamics, energy can never be created

or destroyed within this physical universe~ Energy is identified and measured in many forms, such as thermal, kinetic, nuclear, magnetic, gravitational, elastic, radiant, even mental and emotional~ In Blindsight context, energy is a fundamental active force activated by intention and is infused within every action, object, image, event, and thought~ Prior to the infusion of intention, energy lies dormant in a state of infinite XPotentiality~

5) Entanglement—Blindsight refers to Quantum Entanglement as a proven physical phenomenon occurring when pairs or groups of particles interact so that quantum states of a single particle, such as momentum, position, polarity, etc., cannot be independently described except in respect to the whole system~ Oversimplified, entanglement implies that everything in this universe was inextricably connected at the Big Bang~ There can be no duality in a closed system such as this~ Living as an entangled Energy Signature, there can be no good versus evil, right versus wrong, you versus me, heaven versus hell or ___ versus ___~ This point of view flies in the faces of most of the world's religions, which would have followers believe in a supreme being outside of one's self~ However, if pondered, it becomes most liberating~ From this perspective, for example, if we were to experience the act of hurting someone, whether physically, mentally, or otherwise, we would only be hurting ourselves~ We are as much that person as we are a galaxy or a microbe~ This is likened to the Hindu concept of Karma or Christianity's Golden Rule: "Do unto others as you would have them do unto you." Existing entangled is

to experience unconditional love with unlimited creative XPotential~

~~~*~~~

For your consideration, do you think that had Jesus, Mohammed, Buddha, Abraham, or Krishna been able to view themselves as intelligent energy in a Multiverse of entanglement and superposition, they would have expressed the awareness and experience of their beings in the same language? Blindsight endeavors to understand and express the very same concepts as these greatest of meta-spiritual Intellergies~ The goal is simply Truth~

~~~*~~~

If you believe that you are NOT omnipresent, omniscient, and ultimately omnipotent — you are delusional~
— Kevin Michel

6) Energy Signature—An Energy Signature is a self-aware state of energy and intelligent information~ The addition of *Signature* implies a more defined quality, but should not be construed as consisting of any fixed form~ In our Blindsight discussions, an Energy Signature is most often a human being manifesting from a Qen4 perspective; however, all thoughts, events, and phenomena in the Multiverse are products of energy and intelligent information, so they may also be considered Energy Signatures~

7) Infinity—described as something *without any bound, larger than any number~* This word, along with *infinite, eternal, eternity, sempiternal, unending, endless, limitless, unlimited,* and other such references, appear repeatedly throughout this simulation~ If one has not contemplated these concepts, the present would be the optimum time to take pause and do so, since the above renderings would not carry anywhere near the same Intellergy without at least some attempt at an understanding of a concept that has no end~

8) Intellergy—Intellergy is defined as energy infused with intelligent information in the Cosmic Inflation; also known as the Inflationary Epoch, following the Big Bang~ Information implies intelligence either organized *proactively* by self-aware intention or *reactively* as a result of interaction with an outside force or intention~ For example, a galaxy formed as a result of the Big Bang would be an Intellergy *reaction~* The present, self-aware intentional choice being made to write this sentence is a *proactive* event~ Intellergy can be found in a being, an object, an action, a thought, or an occurrence~

9) Intention—Intention is a mental state that implies forethought and planning from a position of self-awareness~ Intention most often precedes an action taken in pursuit of a goal or outcome~ Intention can arise from a belief system leading to the satisfaction of a desire~ The Buddha would offer that all intention is rooted in desire~ Intention also occurs as a reaction to an event such as an encounter, a threat, or physical contact~ In Blindsight context, intention, whether reactive or proactive, is the essential activating force of creation~

10) Mass — Blindsight recognizes that the mass of elementary particles including electrons, quarks, and the rest are the results of the transfer of potential energy when they interact with the Higgs Field~ The concept of *mass* is very simply a misconception~ In Big Bang theory, there had always been a problem including particles that had mass, but when physicists ran the math without mass, the theory performed perfectly~ So long, reality!

The Multiverse became a big magic trick, an illusion of substance~ Peter Higgs and his team proved the existence of the Higgs Field and its complimentary *discovery* of the Higgs boson~ It can be described as a *crystal-like* field that pervades all space and gives particles the appearance and characteristics of mass, when they are truly made up of Intellergy or intelligent energy~

11) The Observer — In quantum mechanics, observation is also measurement~ Whenever an object, event, or phenomenon is observed or measured, its wave function collapses and it no longer exists as a probability cloud with infinite potential~

The observer, when he seems to be observing a stone, if physics is to be believed, is observing the effects of the stone upon himself~
— Bertrand Russell

Quantum physics tells us that nothing that is observed is unaffected by the observer~ That statement, from science, holds an enormous and powerful insight~ It means

*that everyone sees a different truth, because everyone is
creating what they see~*
— Neale Donald Walsch

Blindsight recognizes there is an *Observer* observing the observation~ This *Observer,* in fact, makes possible the act of observation~ The *Observer* is _____~

12) Quantum Point—A convergence of high densities of energy and information, resulting in a completely new state of form, content, or consciousness~ Commonplace examples might include the boiling of water, converting it into steam, or perhaps the first epiphany that led to the invention of the wheel~

13) Qen (pronounced kehn)—In Blindsight, Qen is ultimately indefinable, and yet a term must be conjured here for the expansion purposes of this simulation~ Qen is the infinite field, pool, continuum, medium from which all manifestation emerges~ Qen is the undisturbed *palette* from which _____ mixes with intention the many *colors* of creation~ Qen may Intellergetically appear as a spatial aspect when observed as such, but Qen may also be perceived as particulate with characteristics simulating mass~ Qen for this Multiverse exists as a Qentinuum, within which the Multiverse evolves~ This is to include all manifestation, Intellergy, and Spacetime~ The closest analogy is the Quantum *dimension;* however, all dimensions evolve within Qen~ Qen assumes an infinite spectrum of frequency expansion when spanned across a Multiverse~ This spectrum has responses reflecting *strata* of harmonics and resonances like unto an FM radio

station at 101.5 MHz that samples from the FM radio band between 88 and 108 MHz~ These strata, although completely entangled with the full spectrum, resonate with *bands* of frequencies that take on multiversal properties loosely individual to the respective sample of the spectrum~ Much like the radio station, these bands assume densities for perception resonance that become Intellergetic in structure~ Qen, in this context, may be assigned harmonics such as Qen4 or Qen5, affording an Energy Signature immersion access and/or experience for transition movement through Thojourn, meditation, hallucinogen, or some other form of proactive Intention~

14) Blindsight — The discussion of consciousness and energy in relation to Quantum Mechanical principles and ancient contemplative philosophies~

15) Signome — A fully entangled, present, and self-actualized Energy Signature Being immersed in _____~

16) Singularity — A quantum position of infinite potential~ Mass, Time, and Space do not exist in a Singularity~ It is from a singularity that the Big Bang emerged~ It has also been attributed to the center of Black Holes; however, we will not be discussing that particular reference here~ A singularity is related to the Qen1 stratum~ It has no height, no width, no depth, and no movement through time, yet it has existence~ It also represents infinite potential of energy and consciousness~

17) Terra Signature — The Global Energy Signature recognizing Earth as possessing Intellergetic aspects~

18) Terranome — A term to describe a thought experiment allowing for a single, fully entangled, self-aware cosmological being at one with _____ in global satori~

19) Thojourn — The word, *sojourn,* used as a verb means to go somewhere for a short time, and it implies intention as in a plan or a map~ As a noun, a sojourn is a temporary stay~ In that context, we offer Thojourn as both noun and verb as a new way of expressing the idea of a mind experiment or thought journey~ Thojourn meditations and creative visualizations are conscious, intentional acts differing from reverie or day dreaming, which are reactive in nature~

20) One Mind — For Blindsight purposes, the all-pervasive, all-knowing manifestation of Intellergy that sustains and evolves this Multiverse~ It is manifested from the intention of _____ that was injected into infinite pools of energy and consciousness at the Big Bang~ It is its permeation throughout all multiversal existence that makes self-aware intelligence possible~ One Mind, in this Multiverse, is a dualistic system: tension and release, light and shadow, expansion and contraction~ Although positive and negative Intellergy exists in equal quantities, the infusion of the power of intention allows for proactive creativity and evolution by Intellergetic beings~

21) XPotential — A state of exponential potential that manifests in an Energy Signature that is fully present in *NOW* without attachment to density or time, similar to the *Zen Moment* described by blindfolded Buddhist archers who send arrow after arrow into the center of moving targets by virtue of their presence, their focus,

and their intention~ Intellergetic humans in this state are capable of extraordinary perceptions and actions~

About the Author

Photo by Alice Rabbit (alicerabbit.com)

Mikal Masters lives with love mate and life partner, G, outside of Austin, Texas~ To hear or purchase Mikal's music, go to Tortugans.com~

www.blindsight.mobi

NOTES

NOTES

NOTES

www.ingramcontent.com/pod-product-compliance
Lightning Source LLC
Chambersburg PA
CBHW060318200326
41519CB00011BA/1766